Seventy-six Stata Tips

Second Edition

Seventy-six Stata Tips

Second Edition

H. JOSEPH NEWTON, Editor
Texas A & M University
Department of Statistics

NICHOLAS J. COX, Editor
Durham University
Department of Geography

A Stata Press Publication
StataCorp LP
College Station, Texas

Contents

Editors' preface   viii

Introducing Stata tips 1
Stata tip 1: The eform() option of regress . . . . R. Newson   2
Stata tip 2: Building with floors and ceilings . . . . N. J. Cox   3
Stata tip 3: How to be assertive . . . . W. Gould   5
Stata tip 4: Using display as an online calculator . . . . P. Ryan   6
Stata tip 5: Ensuring programs preserve dataset sort order . . . . R. Newson   7
Stata tip 6: Inserting awkward characters in the plot . . . . N. J. Cox   8
Stata tip 7: Copying and pasting under Windows . . . . . . S. Driver and P. Royston   10
Stata tip 8: Splitting time-span records with categorical time-varying covariates
. . . . B. Jann   11
Stata tip 9: Following special sequences . . . . N. J. Cox   13
Stata tip 10: Fine control of axis title positions . . . . P. Ryan and N. Winter   14
Stata tip 11: The nolog option with maximum-likelihood modeling commands . .
. . . . P. Royston   16
Stata tip 12: Tuning the plot region aspect ratio . . . . N. J. Cox   17
Stata tip 13: generate and replace use the current sort order . . . . R. Newson   19
Stata tip 14: Using value labels in expressions . . . . K. Higbee   21
Stata tip 15: Function graphs on the fly . . . . N. J. Cox   23
Stata tip 16: Using input to generate variables . . . . U. Kohler   25
Stata tip 17: Filling in the gaps . . . . N. J. Cox   26
Stata tip 18: Making keys functional . . . . S. Driver   28
Stata tip 19: A way to leaner, faster graphs . . . . P. Royston   30
Stata tip 20: Generating histogram bin variables . . . . D. A. Harrison   31
Stata tip 21: The arrows of outrageous fortune . . . . N. J. Cox   33
Stata tip 22: Variable name abbreviation . . . . P. Ryan   36
Stata tip 23: Regaining control over axis ranges . . . . N. Winter   38
Stata tip 24: Axis labels on two or more levels . . . . N. J. Cox   40
Stata tip 25: Sequence index plots . . . . U. Kohler and C. Brzinsky-Fay   41
Stata tip 26: Maximizing compatibility between Macintosh and Windows . . . .
. . . . M. S. Hanson   43
Stata tip 27: Classifying data points on scatter plots . . . . N. J. Cox   44

Stata tip 28: Precise control of dataset sort order...................P. Schumm   47

Stata tip 29: For all times and all places .... C. H. Franklin   50

Stata tip 30: May the source be with you .... N. J. Cox   52

Stata tip 31: Scalar or variable? The problem of ambiguous names .. G. I. Kolev   54

Stata tip 32: Do not stop..S. P. Jenkins   56

Stata tip 33: Sweet sixteen: Hexadecimal formats and precision problems.......
....N. J. Cox   57

Stata tip 34: Tabulation by listing..............................D. A. Harrison   59

Stata tip 35: Detecting whether data have changed .... W. Gould   62

Stata tip 36: Which observations? .... N. J. Cox   64

Stata tip 37: And the last shall be first .... C. F. Baum   67

Stata tip 38: Testing for groupwise heteroskedasticity...............C. F. Baum   69

Stata tip 39: In a list or out? In a range or out? .... N. J. Cox   72

Stata tip 36: Which observations? Erratum..........................N. J. Cox   75

Stata tip 40: Taking care of business...............................C. F. Baum   76

Stata tip 41: Monitoring loop iterations .... D. A. Harrison   79

Stata tip 42: The overlay problem: Offset for clarity.....................J. Cui   80

Stata tip 43: Remainders, selections, sequences, extractions: Uses of the modulus
....N. J. Cox   82

Stata tip 44: Get a handle on your sample............................B. Jann   85

Stata tip 45: Getting those data into shape .... C. F. Baum and N. J. Cox   87

Stata tip 46: Step we gaily, on we go .... R. Williams   91

Stata tip 47: Quantile–quantile plots without programming...........N. J. Cox   94

Stata tip 48: Discrete uses for uniform() .... M. L. Buis   99

Stata tip 49: Range frame plots .... S. Merryman   101

Stata tip 50: Efficient use of summarize..............................N. J. Cox   103

Stata tip 51: Events in intervals .... N. J. Cox   105

Stata tip 52: Generating composite categorical variables .... N. J. Cox   109

Stata tip 53: Where did my p-values go?...........................M. L. Buis   111

Stata tip 54: Post your results....................................P. Van Kerm   114

Stata tip 55: Better axis labeling for time points and time intervals ... N. J. Cox   117

Stata tip 56: Writing parameterized text files .... R. Gini   120

Stata tip 57: How to reinstall Stata................................W. Gould   123

Stata tip 58: nl is not just for nonlinear models.......................B. P. Poi   125

Stata tip 59: Plotting on any transformed scale .... N. J. Cox   128

Stata tip 60: Making fast and easy changes to files with filefilter.....A. R. Riley   132

Stata tip 61: Decimal commas in results output and data input .... N. J. Cox   135

Stata tip 62: Plotting on reversed scales........N. J. Cox and N. L. M. Barlow   137

Stata tip 63: Modeling proportions .... C. F. Baum   141

Stata tip 64: Cleaning up user-entered string variables....J. Herrin and E. Poen   146

Stata tip 65: Beware the backstabbing backslash......................N. J. Cox   148

Stata tip 66: ds—A hidden gem.......................................M. Weiss   150

Stata tip 67: J() now has greater replicating powers..................N. J. Cox   152

Stata tip 69: Producing log files based on successful interactive commands......
.... A. R. Riley   154

Stata tip 70: Beware the evaluating equal sign.......................N. J. Cox   157

Stata tip 71: The problem of split identity, or how to group dyads .... N. J. Cox   159

Stata tip 72: Using the Graph Recorder to create a pseudograph scheme........
.... K. Crow   163

Stata tip 73: append with care!....................................C. F. Baum   165

Stata tip 74: firstonly, a new option for tab2...................................
.... R. G. Gutierrez and P. A. Lachenbruch   168

Stata tip 75: Setting up Stata for a presentation ....K. Crow   170

Stata tip 76: Separating seasonal time series........................N. J. Cox   172

Editors' preface

The book you are reading reprints 76 Stata Tips from the *Stata Journal*, with thanks to their original authors. We, the *Journal* editors, began publishing tips in 2003, beginning with volume 3, number 4. It pleases us now to reprint them in this book.

The *Stata Journal* publishes substantive and peer-reviewed articles ranging from reports of original work to tutorials on statistical methods and models implemented in Stata, and indeed on Stata itself. The original material we have published since 2001 includes special issues such as those on measurement error models (volume 3, number 4, 2003) and simulated maximum likelihood (volume 6, number 2, 2006).

Other features include regular columns on Stata (currently, "Speaking Stata" and "Mata Matters"), book reviews, and announcements of software updates.

We are pleased by the external recognition that the *Journal* has achieved. In 2005, it was added to two of Thomson Scientific's citation indexes, the Science Citation Index Expanded and the CompuMath Citation Index.

But back to the Tips. There was little need for tips in the early days. Stata 1.0 was released in 1985. The original program had 44 commands, and its documentation totaled 175 pages. Stata 11, on the other hand, has more than 950 commands—including an embedded matrix language called Mata—and Stata's official documentation now totals more than 8,300 pages. Beyond that, the user community has added several hundred more commands.

The pluses and the minuses of this growth are evident. As Stata expands, it is increasingly likely that users' needs can be met by available code. But at the same time, learning how to use Stata and even learning what is available become larger and larger tasks.

The Tips are intended to help. The ground rules for Stata Tips, as found in the original 2003 statement, are laid out as the next item in this book. We have violated one original rule in the letter, if not the spirit: Some Stata Tips have been as long as six pages. However, the intention of producing concise tips that are easy to pick up remains as it was.

The Tips grew from many discussions and postings on Statalist, at Users Group meetings, and elsewhere, which underscores a simple fact: Stata is now so big that it is easy to miss even simple features that can streamline and enhance your sessions with Stata. This applies not just to new users, who understandably may quake nervously before the manual mountain, but also to longtime users, who too are faced with a mass of new features in every release.

Tips have come from Stata users as well as from StataCorp employees. Many discuss new features of Stata, or features not documented fully or even at all. We hope that you enjoy the Stata Tips reprinted here and can share them with your fellow Stata users. If you have tips that you would like to write, or comments on the kinds of tips that are helpful, do get in touch with us, as we are eager to continue the series.

H. Joseph Newton, Editor
Nicholas J. Cox, Editor

The Stata Journal (2003)
3, Number 4, p. 328

1

Introducing Stata tips

As promised in our editorial in *Stata Journal* 3(2), 105–108 (2003), the *Stata Journal* is hereby starting a regular column of tips. Stata tips will be a series of concise notes about Stata commands, features, or tricks that you may not yet have encountered.

The examples in this issue should indicate the kinds of tips we will publish. What we most hope for is that readers are left feeling, "I wish I had known that earlier!" Beyond that, here are some more precise guidelines:

Content A tip will draw attention to useful details in Stata or in the use of Stata. We are especially keen to publish tips of practical value to a wide range of users. A tip could concern statistics, data management, graphics, or any other use of Stata. It may include advice on the user interface or about interacting with the operating system. Tips may explain pitfalls (do not do this) as well as positive features (do use this). Tips will not include plugs for user-written programs, however smart or useful.

Length Tips must be brief. A tip will take up at most three printed pages. Often a code example will explain just as much as a verbal discussion.

Authorship We welcome submissions of tips from readers. We also welcome suggestions of tips or of kinds of tips you would like to see, even if you do not feel that you are the person to write them. Naturally, we also welcome feedback on what has been published. An email to *editors@stata-journal.com* will reach us both.

H. Joseph Newton, Editor
Texas A&M University
jnewton@stat.tamu.edu

Nicholas J. Cox, Editor
University of Durham
n.j.cox@durham.ac.uk

Stata tip 1: The eform() option of regress

Roger Newson, King's College London, UK
roger.newson@kcl.ac.uk

Did you know about the `eform()` option of `regress`? It is very useful for calculating confidence intervals for geometric means and their ratios. These are frequently used with skewed Y-variables, such as house prices and serum viral loads in HIV patients, as approximations for medians and their ratios. In Stata, I usually do this by using the `regress` command on the logs of the Y-values, with the `eform()` and `noconstant` options. For instance, in the `auto` dataset, we might compare prices between non-US and US cars as follows:

```
. sysuse auto, clear
(1978 Automobile Data)

. generate logprice = log(price)

. generate byte baseline = 1

. regress logprice foreign baseline, noconstant eform(GM/Ratio) robust
Regression with robust standard errors                Number of obs =      74
                                                      F(  2,    72) =18043.56
                                                      Prob > F      =  0.0000
                                                      R-squared     =  0.9980
                                                      Root MSE      = .39332
```

logprice	GM/Ratio	Robust Std. Err.	t	P>|t|	[95% Conf. Interval]	
foreign	1.07697	.103165	0.77	0.441	.8897576	1.303573
baseline	5533.565	310.8747	153.41	0.000	4947.289	6189.316

We see from the `baseline` parameter that US-made cars had a geometric mean price of 5534 dollars (95% CI from 4947 to 6189 dollars), and we see from the `foreign` parameter that non-US cars were 108% as expensive (95% CI, 89% to 130% as expensive). An important point is that, if you want to see the baseline geometric mean, then you must define the constant variable, here `baseline`, and enter it into the model with the `noconstant` option. Stata usually suppresses the display of the intercept when we specify the `eform()` option, and this trick will fool Stata into thinking that there is no intercept for it to hide. The same trick can be used with `logit` using the `or` option, if you want to see the baseline odds as well as the odds ratios.

My nonstatistical colleagues understand regression models for log-transformed data a lot better this way than any other way. Continuous X-variables can also be included, in which case the parameter for each X-variable is a ratio of Y-values per unit change in X, assuming an exponential relationship—or assuming a power relationship, if X is itself log-transformed.

The Stata Journal (2003)
3, Number 4, pp. 446–447

Stata tip 2: Building with floors and ceilings

Nicholas J. Cox, University of Durham, UK
n.j.cox@durham.ac.uk

Did you know about the `floor()` and `ceil()` functions added in Stata 8?

Suppose that you want to round down in multiples of some fixed number. For concreteness, say, you want to round `mpg` in the auto data in multiples of 5 so that any values 10–14 get rounded to 10, any values 15–19 to 15, etc. `mpg` is simple, in that only integer values occur; in many other cases, we clearly have fractional parts to think about as well.

Here is an easy solution: `5 * floor(mpg/5)`. `floor()` always rounds down to the integer less than or equal to its argument. The name floor is due to Iverson (1962), the principal architect of APL, who also suggested the expressive $\lfloor x \rfloor$ notation. For further discussion, see Knuth (1997, 39) or Graham, Knuth, and Patashnik (1994, chapter 3).

As it happens, `5 * int(mpg/5)` gives exactly the same result for `mpg` in the auto data, but in general, whenever variables may be negative as well as positive, *interval* * `floor(`*expression/interval*`)` gives a more consistent classification.

Let us compare this briefly with other possible solutions. `round(mpg, 5)` is different, as this rounds to the nearest multiple of 5, which could be either rounding up or rounding down. `round(mpg - 2.5, 5)` should be fine but is also a little too much like a dodge.

With `recode()`, you need two dodges, say, `-recode(-mpg,-40,-35,-30,-25,-20,` `-15,-10)`. Note all the negative signs; negating and then negating to reverse it are necessary because `recode()` uses its numeric arguments as upper limits; i.e., it rounds up.

`egen, cut()` offers another solution with option call `at(10(5)45)`. Being able to specify a *numlist* is nice, as compared with spelling out a comma-separated list, but you *must* also add a limit, here 45, which will not be used; otherwise, with `at(10(5)40)`, your highest class will be missing.

Yutaka Aoki also suggested to me `mpg - mod(mpg,5)`, which follows immediately once you see that rounding down amounts to subtracting the appropriate remainder. `mod(,)`, however, does not offer a correspondingly neat way of rounding up.

The `floor` solution grows on one, and it has the merit that you do not need to spell out all the possible end values, with the risk of forgetting or mistyping some. Conversely, `recode()` and `egen, cut()` are not restricted to rounding in equal intervals and remain useful for more complicated problems.

Without recapitulating the whole argument insofar as it applies to rounding up, `floor()`'s sibling `ceil()` (short for ceiling) gives a nice way of rounding up in equal intervals and is easier to work with than expressions based on `int()`.

References

Graham, R. L., D. E. Knuth, and O. Patashnik. 1994. *Concrete Mathematics: A Foundation for Computer Science.* Reading, MA: Addison–Wesley.

Iverson, K. E. 1962. *A Programming Language.* New York: Wiley.

Knuth, D. E. 1997. *The Art of Computer Programming. Volume 1: Fundamental Algorithms.* Reading, MA: Addison–Wesley.

The Stata Journal (2003)
3, Number 4, p. 448

Stata tip 3: How to be assertive

William Gould, StataCorp
wgould@stata.com

`assert` verifies the truth of a claim:

```
. assert sex=="m" | sex=="f"
. assert age>=18 & age<=65
22 contradictions in 2740 observations
assertion is false
r(9);
```

The best feature of `assert` is that, when the claim is false, it stops do-files and ado-files:

```
. do my_data_prep
. use basedata, clear
. assert age>=18 & age<=64
22 contradictions in 2740 observations
assertion is false
r(9);
end of do-file
r(9);
```

`assert` has two main uses:

1. It checks that claims made to you and suppositions you have made about the data you are about to process are true:

   ```
   . assert exp==. if age<18
   . assert exp<. if age>=18
   ```

2. It tests that, when you write complicated code, the code produces what you expect:

   ```
   . sort group
   . by group: gen avg = sum(hours)/sum(hours<.)
   . by group: assert avg!=. if _n==_N
   . by group: gen relative = hours/avg[_N]
   ```

`assert` is especially useful following `merge`:

```
. merge id
. sort id using demog
. assert _merge==3
. drop _merge
```

The Stata Journal (2004)
4, Number 1, p. 93

Stata tip 4: Using display as an online calculator

Philip Ryan, University of Adelaide
philip.ryan@adelaide.edu.au

Do you use Stata for your data management, graphics, and statistical analysis but switch to a separate device for quick calculations? If so, you might consider the advantages of using Stata's built-in `display` command:

1. It is always at hand on your computer.

2. As with all Stata calculations, double precision is used.

3. You can specify the format of results.

4. It uses and reinforces your grasp of Stata's full set of built-in functions.

5. You can keep an audit trail of results and the operations that produced those results, as part of a log file. You can also add extra comments to the output.

6. Editing of complex expressions is easy, without having to re-enter lengthy expressions after a typo.

7. You can copy and paste results elsewhere whenever your platform supports that.

8. It is available via the menu interface (select **Data—Other utilities—Hand calculator**).

9. It can be abbreviated to `di`.

To be fair, there are some disadvantages, such as its lack of support for Reverse Polish Notation or complex number arithmetic, but in total, `display` provides you with a powerful but easy-to-use calculator.

```
. di _pi
3.1415927
. di %12.10f _pi
3.1415926536
. * probability of 2 heads in 6 tosses of a fair coin
. di comb(6,2) * 0.5^2 * 0.5^4
.234375
. di "chi-square (1 df) cutting off 5% in upper tail is " invchi2tail(1, .05)
chi-square (1 df) cutting off 5% in upper tail is 3.8414588
. * Euler-Mascheroni gamma
. di %12.10f -digamma(1)
0.5772156649
```

The Stata Journal (2004)
4, Number 1, p. 94

Stata tip 5: Ensuring programs preserve dataset sort order

Roger Newson, King's College London, UK
roger.newson@kcl.ac.uk

Did you know about `sortpreserve`? If you are writing a Stata program that temporarily changes the order of the data and you want the data to be sorted in its original order at the end of execution, you can save a bit of programming by including `sortpreserve` on your `program` statement. If your program is called `myprogram`, you can start it with

```
program myprogram, sortpreserve
```

If you do this, you can change the order of observations in the dataset in `myprogram`, and Stata will automatically sort it in its original order at the end of execution. Stata does this by creating a temporary variable whose name is stored in a macro named `_sortindex`, which is discussed in the manuals under [P] **sortpreserve**. (Note, however, that there is a typo in the manual; the underscore in `_sortindex` is missing.[1]) The temporary variable '`_sortindex`' contains the original sort order of the data, and the dataset is sorted automatically by '`_sortindex`' at the end of the program's execution.

If you know about temporary variables, you might think that `sortpreserve` is unnecessary because you can always include two lines at the beginning, such as

```
tempvar order
generate long 'order' = _n
```

and a single line at the end such as

```
sort 'order'
```

and do the job of `sortpreserve` in 3 lines. However, `sortpreserve` does more than that. It restores the result of the macro extended function `sortedby` to the value that it would have had before your program executed. (See [P] **macro** for a description of `sortedby`.) Also, it restores the "Sorted by:" variable list reported by the `describe` command to the variable list that would have been reported before your program executed. For example, in the `auto` dataset shipped with official Stata, the output of `describe` ends with the message

```
Sorted by:  foreign
```

This will not be changed if you execute a program defined with `sortpreserve`.

---

1. This has been fixed in Stata 10 and later manuals.

The Stata Journal (2004)
4, Number 1, pp. 95–96

Stata tip 6: Inserting awkward characters in the plot[1]

Nicholas J. Cox, University of Durham, UK
n.j.cox@durham.ac.uk

Did you know about the function `char()`? `char(n)` returns the character corresponding to ASCII code n for $1 \leq n \leq 255$. There are several numbering schemes for so-called ASCII characters. Stata uses the ANSI scheme; a web search for "ANSI character set" will produce tables showing available characters. This may sound like an arcane programmer's tool, but it offers a way to use awkward text characters—either those not available through your keyboard or those otherwise problematic in Stata. A key proviso, however, is that you must have such characters available in the font that you intend to use. Fonts available tend to vary not only with platform but even down to what is installed on your own system. Some good fonts for graphics, in particular, are Arial and Times New Roman.

Let us see how this works by considering the problem of inserting awkward characters in your Stata graphs, say as part of some plot or axis title. Some examples of possibly useful characters are

`char(133)`	ellipsis	…
`char(134)`	dagger	†
`char(135)`	double dagger	‡
`char(169)`	copyright	©
`char(176)`	degree symbol	°
`char(177)`	plus or minus	±
`char(178)`	superscript 2	^{2}
`char(179)`	superscript 3	^{3}
`char(181)`	micro symbol	μ
`char(188)`	one-fourth	$\frac{1}{4}$
`char(189)`	one-half	$\frac{1}{2}$
`char(190)`	three-fourths	$\frac{3}{4}$
`char(215)`	multiply	×

There are many others that might be useful to you, including a large selection of accented letters of the alphabet. You can use such characters indirectly or directly. The indirect way is to place such characters in a local macro and then to refer to that macro within the same program or do-file.

For example, I use data on river discharge, for which the standard units are cubic meters per second. I can get the cube power in an axis title like this:

---

1. Editors' note: Stata 11 introduced a much easier way to add symbols and Greek letters to graph text. See [G] *text* or `help graph text`.

```
. local cube = char(179)
. scatter whatever, xtitle("discharge, m`cube'/s")
```

Or, I have used Hanning, a binomial filter of length 3:

```
. local half = char(189)
. local quarter = char(188)
. twoway connected whatever,
>         title("Smoothing with weights `quarter':`half':`quarter'")
```

The direct way is to get a macro evaluation on the fly. You can write `` `=char(176)' ``
and, in one step, get the degree symbol (for temperatures or compass bearings). This
feature was introduced in Stata 7 but not documented until Stata 8. See [P] **macro**.

`char()` has many other uses besides graphs. Suppose that a string variable contains
fields separated by tabs. For example, `insheet` leaves tabs unchanged. Knowing that
a tab is `char(9)`, we can

```
. split data, p(`=char(9)') destring
```

Note that `p(char(9))` would not work. The argument to the `parse()` option is
taken literally, but the function is evaluated on the fly as part of macro substitution.

Note also that the SMCL directive `{c #}` may be used for some but not all of these
purposes. See [P] **smcl**. Thus,

```
. scatter whatever, xtitle("discharge, m{c 179}/s")
```

would work, but using a SMCL directive would not work as desired with `split`.

The Stata Journal (2004)
4, Number 2, p. 220

Stata tip 7: Copying and pasting under Windows

Shannon Driver
StataCorp
sdriver@stata.com

Patrick Royston
MRC Clinical Trials Unit, London
patrick.royston@ctu.mrc.ac.uk

Windows users often copy, cut, and paste material between applications or between windows within applications. Here are two ways you can do this with Stata for Windows. We will describe one as a mouse-and-keyboard operation and the other as a menu-based operation. Experienced Windows users will know that these methods are, to a large extent, alternatives.

First, you can highlight some text in the Results window, copy it using the mouse (or keyboard), and then paste it into the Command window, the Do-file Editor, or anywhere else appropriate. This is a convenient way to transfer, for example, single values, lists, or sets of variable names from the screen for use in the next command. To copy text, place your mouse at the beginning of the desired text, drag to the end, thus highlighting the selected text, and press Ctrl-C. To paste text, click your mouse at the appropriate place and press Ctrl-V.

Suppose that a local macro 'macro' holds some text you wish to use. Then type

```
. display "'macro'"
```

and copy and paste the contents of 'macro' for editing in the Command window. Or, list in alphabetic order the names of variables not beginning with _I:

```
. ds _I*, not alpha
```

and then copy and paste the list into the Do-file Editor.

Second, suppose that you want to save a table constructed using `tabstat` in a form that makes it easy to convert into a table in MS Word. Stata has a **Copy Table** feature that you might find very useful. Make sure at the outset that you have set suitable options by clicking **Edit** in the menu bar and then **Table Copy Options**. In this case, removing all the vertical bars is advisable, so make sure **Remove all** is selected, and click **OK**. Now highlight the table in the Results window, and then click **Edit** and then **Copy Table**.

In MS Word, click **Edit** and then **Paste**. Highlight the pasted text and then click **Table** and then **Convert** and **Text to Table**. Specify **Tabs** under the **Separate text at** if it is not already selected. Click **OK** to create your table.

The Stata Journal (2004)
4, Number 2, pp. 221–222

Stata tip 8: Splitting time-span records with categorical time-varying covariates

Ben Jann, ETH Zürich, Switzerland
jann@soz.gess.ethz.ch

In survival analysis, time-varying covariates are often handled by the method of episode splitting. The stsplit command does this procedure very well, especially in the case of continuous time-varying variables such as age or time in study. Quite often, however, we are interested in evaluating the effect of a change in some kind of categorical status or the occurrence of some secondary event. For example, we might be interested in the effect of the birth of a child on the risk of divorce or the effect of having completed further education on the chances of upward occupational mobility.

In such situations, the creation of splits might appear to be more complicated, and stsplit does not seem to be of much help, at least judging from the rather complicated examples provided with [ST] **stset** (*Final example: Stanford heart transplant data*) and [ST] **stsplit** (*Example 3: Explanatory variables that change with time*). Fortunately, the procedure is simpler than it appears.

Consider the Stanford heart transplant data used in examples for [ST] **stset** and [ST] **stsplit**:

```
. use http://www.stata-press.com/data/r8/stanford, clear
(Heart transplant data)
. list id transplant wait stime died if id==44 | id==16
```

	id	transp~t	wait	stime	died
33.	44	0	0	40	1
34.	16	1	20	43	1

The goal here is to split the single time-span records into episodes before and after transplantation (e.g., to split case 16 at time 20). This can easily be achieved by splitting "at 0" "after wait", the time of transplantation. Note that, if no transplantation was carried out at all, wait should be recoded to a value larger than the observed maximum episode duration (maximum of stime) before the stsplit command is applied:

```
. replace wait = 10000 if wait == 0
(34 real changes made)
. stset stime, failure(died) id(id)
  (output omitted)
. stsplit posttran, after(wait) at(0)
(69 observations (episodes) created)
. replace posttran = posttran + 1
(172 real changes made)
```

```
. list id _t0 _t posttran if id == 44 | id == 16
```

	id	_t0	_t	posttran
23.	16	0	20	0
24.	16	20	43	1
70.	44	0	40	0

It is now possible to evaluate the effect of transplantation on survival time using streg, for example, or to plot survivor functions with time-dependent group membership:

```
. sts graph, by(posttran)

        failure _d:  died
  analysis time _t:  stime
               id:  id
```

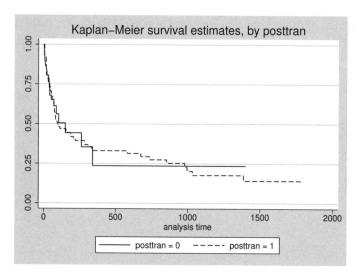

Note that the situation is even simpler if a Cox proportional hazards model is to be fitted. As explained in [ST] **stsplit**, the partial likelihood estimator takes only the times at which failures occur into account. Thus, in the context of Cox regression, the following code would do:

```
. use http://www.stata-press.com/data/r8/stanford, clear
. stset stime, failure(died) id(id)
. stsplit, at(failures)
. generate posttran = wait<_t & wait!=0
. stjoin
. stcox age posttran surgery year
```

The Stata Journal (2004)
4, Number 2, p. 223

Stata tip 9: Following special sequences

Nicholas J. Cox
University of Durham, UK
n.j.cox@durham.ac.uk

Did you know about the special sequences stored as c-class values? [P] **creturn** documents various constant and current values, which may be seen by `creturn list` or which may be accessed once you know their individual names. For example, `c(filename)` stores the name of the file last specified with a `use` or `save` in the current session. However, various special sequences have been added in updates on 1 July 2003 and 15 December 2003 and so are not documented in the manuals. Here is the list:

- `c(alpha)` returns a string containing a space-separated list of the lowercase letters.

- `c(ALPHA)` returns a string containing a space-separated list of the uppercase letters.

- `c(Mons)` returns a string containing a space-separated list of month names abbreviated to three characters.

- `c(Months)` returns a string containing a space-separated list of month names.

- `c(Wdays)` returns a string containing a space-separated list of weekday names abbreviated to three characters.

- `c(Weekdays)` returns a string containing a space-separated list of weekday names.

Even the display of one of these lists can be useful. Note the local macro notation ' ' ensuring that the contents of the list are shown, not its name:

```
. display "`c(Months)'"
```

A common application of these lists is specifying variable or value labels. Suppose that a variable `month` included values 1 to 12. We might type

```
. tokenize `c(Months)'
. forvalues i = 1/12 {
  2. label def month `i' "``i''" , modify
  3. }
. label val month month
```

Finally, the SSC archive (see [R] **ssc**) is organized alphabetically using folders `a` through `z` and `_`. We could get a complete listing of what was available by

```
. foreach l in `c(alpha)' _ {
  2. ssc describe `l'
  3. }
```

The Stata Journal (2004)
4, Number 3, pp. 354–355

Stata tip 10: Fine control of axis title positions

Philip Ryan
University of Adelaide
philip.ryan@adelaide.edu.au

Nicholas Winter
Cornell University
nw53@cornell.edu

ytitle(), xtitle(), and other similar options specify the titles that appear on the axes of Stata graphs (see [G] *axis_title_options*). Usually, Stata's default settings produce titles with a satisfactory format and position relative to the axis. Sometimes, however, you will need finer control over position, especially if there is inadequate separation of the title and the numeric axis labels. This might happen, for example, with certain combinations of the font of the axis labels, the angle the labels make with the axis, the length of the labels, and the size of the graph region.

Although the options ylabel() and xlabel() have a suboption labgap() allowing user control of the gap between tick marks and labels (see [G] *axis_label_options*), the axis title options have no such suboption. The flexibility needed is provided by options controlling the textbox that surrounds the axis title (see [G] *textbox_options*). This box is invisible by default but can be displayed using the box suboption on the axis title option:

```
. graph twoway scatter price weight,
        ytitle("Price of Cars in {c S|}US", box)
        ylab(0(1000)15000, angle(horizontal) labsize(medium))
```

(Note the use of a SMCL directive to render the dollar sign; see [P] **smcl**, page 393.) We can manipulate the relative size of the height of the textbox or the margins around the text within the box to induce the appearance of a larger or smaller gap between the axis title and the axis labels. For a larger gap, we might try one of these solutions:

```
. graph twoway scatter price weight,
        ytitle("Price of Cars in {c S|}US", height(10))
        ylab(0(1000)15000, ang(hor) labsize(medium))
. graph twoway scatter price weight,
        ytitle("Price of Cars in {c S|}US", margin(0 10 0 0))
        ylab(0(1000)15000, ang(hor) labsize(medium))
```

For a smaller gap, specify negative arguments, say, height(-1) in the first command or margin(0 -4 0 0) in the second. A bit of trial and error will quickly give a satisfactory result.

Note that a sufficiently large negative argument in either height() or margin() will permit an axis title to be placed within the inner plot region, namely, inside of the axis. However, this, in turn, may cause the axis labels to disappear off the graph, so that some fiddling with the graphregion() option and its own margin() suboption may then be required (see [G] *region_options* and [G] *marginstyle*). For example,

```
. graph twoway scatter price weight,
        ytitle("Price of Cars in {c S|}US", height(-20))
        ylab(0(1000)15000, ang(hor) labsize(medium))
```

```
. graph twoway scatter price weight,
       ytitle("Price of Cars in {c S|}US", height(-20))
       ylab(0(1000)15000, ang(hor) labsize(medium))
       graphregion(margin(l+20))
```

margin() allows more flexibility in axis title positioning than does height(), but
the price is a slightly more complicated syntax. For example, the y axis title may be
moved farther from the axis labels and closer to the top of the graph by specifying both
the right-hand margin and the bottom margin of the text within the box:

```
. graph twoway scatter price weight,
       ytitle("Price of Cars in {c S|}US", margin(0 10 40 0))
       ylab(0(1000)15000, ang(hor) labsize(medium))
```

The Stata Journal (2004)
4, Number 3, p. 356

Stata tip 11: The nolog option with maximum-likelihood modeling commands

Patrick Royston
MRC Clinical Trials Unit, London
patrick.royston@ctu.mrc.ac.uk

Many Stata commands fit a model by maximum likelihood, and in so doing, they include a report on the iterations of the algorithm towards (it is hoped) eventual convergence. There may be tens or even hundreds or thousands of such lines in a report, which are faithfully recorded in any log file you may have open. Suppose that you did this with `stcox`:

```
. use http://www.stata-press.com/data/r8/cancer, clear
. describe
. stset studytime, failure(died)
. xi: stcox i.drug age, nohr
```

You get six useless lines of output detailing the progress of the algorithm. This is a nice example, as sometimes progress is much slower or more complicated.

Those lines are of little or no statistical interest in most examples and may be omitted by adding the `nolog` option:

```
. xi: stcox i.drug age, nohr nolog
```

The `nolog` option works with many vital Stata commands, including `glm`, `logistic`, `streg`, and several more. My own view is that `nolog` should be the default in all of them. Be that as it may, you can compress your logs to good effect by specifying `nolog` routinely. It will remain obvious when your estimation fails to converge.

The Stata Journal (2004)
4, Number 3, pp. 357–358

Stata tip 12: Tuning the plot region aspect ratio

Nicholas J. Cox
University of Durham, UK
n.j.cox@durham.ac.uk

Sometimes you want a graph to have a particular shape. Graph shape is customarily quantified by the aspect ratio (height/width). One standard way of controlling the aspect ratio is setting the graph height and width by specifying the `ysize()` and `xsize()` options of `graph display`. See also [G] **graph display** and [G] *region_options*. These options control the size and, thus, the shape of the entire available graph area, including titles and other stuff beyond the plot region. At best, this is an indirect way of controlling the plot region shape, which is likely to be your main concern.

In the 23 July 2004 update, Stata 8 added an `aspect()` option to `graph` to meet this need. For example, `aspect(1)` specifies equal height and width, so that the rectangular plot region becomes a square. (A rectangle that is not a square is, strictly, an oblong.) You might want a square plot as a matter either of logic or of taste. Suppose that you are contemplating uniform random numbers falling like raindrops on the unit square within the real plane (or the plain, as the old song has it):

```
. clear
. set obs 100
. gen y = uniform()
. gen x = uniform()
. scatter y x, aspect(1) xla(0(0.1)1, grid) yla(0(0.1)1, ang(h) grid)
> yti(, orient(horiz)) plotregion(margin(none))
```

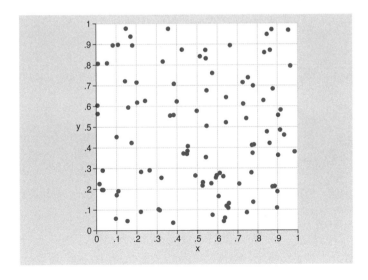

In effect you have drawn a map, and maps customarily have equal vertical and horizontal distance scales or, more simply put, a single distance scale. Hence the aspect ratio is set as 1 whenever the plot region has equal extent on both axes. The same preference applies also to various special graphs on the unit square, such as ROC or Lorenz curves.

In other circumstances, the aspect ratio sought might differ from 1. Fisher (1925, 31) recommended plotting data so that lines make approximately equal angles with both axes; the same advice of banking to 45° is discussed in much more detail by Cleveland (1993). Avoiding roller-coaster plots of time series is one application. In practice, a little trial and error will be needed to balance a desire for equal angles with other considerations. For example, try variations on the following:

```
. use http://www.stata-press.com/data/r8/htourism.dta
. tsline mvdays, aspect(0.15) yla(, ang(h))
```

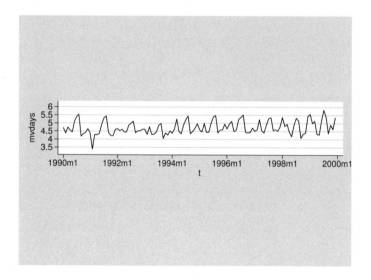

References

Cleveland, W. S. 1993. *Visualizing Data*. Summit, NJ: Hobart Press.

Fisher, R. A. 1925. *Statistical Methods for Research Workers*. Edinburgh: Oliver & Boyd.

The Stata Journal (2004)
4, Number 4, pp. 484–485

Stata tip 13: generate and replace use the current sort order

Roger Newson
King's College London, UK
roger.newson@kcl.ac.uk

Did you know that `generate` and `replace` use the current sort order? You might have guessed this because otherwise the `sum()` function could work as designed only with difficulty. However, this fact is not documented in the manuals, but only in the Stata web site FAQs. The consequence is that, given a particular desired `sort` order, you can be sure that values of a variable are calculated in that order and can use them to calculate subsequent values of the same variable.

A simple example is filling in missing values by copying the previous nonmissing value. The syntax for this is simply

```
. replace myvar = myvar[_n-1] if missing(myvar)
```

Here the subscript `[_n-1]`, based on the built-in variable `_n`, refers to the previous observation in the present sort order. To find more about subscripts, see [U] **13.7 Explicit subscripting** or the online help for `subscripting`.

Suppose that values of `myvar` are present for observations 1, 2, and 5 but missing in observations 3, 4, and 6. `replace` starts by replacing `myvar[3]` with the nonmissing `myvar[2]`. It then replaces `myvar[4]` with `myvar[3]`, which now contains (just in time) a copy of the nonmissing `myvar[2]`. Finally, `replace` puts a copy of `myvar[5]` into `myvar[6]`. As said, this all requires that data are in the desired sort order, commonly that of some time variable. If not, reach for the `sort` command.

There are numerous variations on this idea. Suppose that a sequence of years contains nonmissing values only for years like 1980, 1990, and 2000. This is common in data derived from spreadsheet files. A simple fix would be

```
. replace year = year[_n-1] + 1 if mi(year)
```

That way, changes cascade down the observations.

More exotic examples concern recurrence relations, as found in probability theory and elsewhere in mathematics. We typically use `generate` to define the first value (or the first few values) and `replace` to define the other values.

Consider the famous "birthday problem": what is the probability that no two out of n people have the same birthday? Assuming equal probabilities of birth on each of 365 days, and so ignoring leap years and seasonal fertility variation, this probability is $\prod_{j=1}^{n} x_j$, where $x_j = (365 - j + 1)/365$. We can put these probabilities into a variable `palldiff` by typing

```
. set obs 370
. generate double palldiff = 1
. replace palldiff = palldiff[_n-1] * (365 - _n + 1) / 365 in 2/1
. label var palldiff "Pr(All birthdays are different)"
. list palldiff
```

To illustrate, the probability that all birthdays are different is below 0.5 for 23 people, below one-millionth for 97 people, and zero for over 365 people. An alternative solution (based on a suggestion by Roberto Gutierrez) is to replace the second and third lines of the above program with

```
. generate double palldiff = 0
. replace palldiff = exp(sum(ln(366 - _n) - ln(365))) in 1/365
```

which works because the product of positive numbers is the sum of their logarithms, exponentiated.

Another example is the Fibonacci sequence, defined by $y_1 = y_2 = 1$ and otherwise by $y_n = y_{n-1} + y_{n-2}$. The first 20 numbers are given by

```
. set obs 20
. generate y = 1
. replace y = y[_n-1] + y[_n-2] in 3/1
. list y
```

If you ever want to work backwards by referring to later observations, it is often easiest to reverse the order of observations and then to use tricks like these.

The Stata Journal (2004)
4, Number 4, pp. 486–487

Stata tip 14: Using value labels in expressions

Kenneth Higbee
StataCorp
khigbee@stata.com

Did you know that there is a way in Stata to specify value labels directly in an expression, rather than through the underlying numeric value? You specify the label in double quotes (" "), followed by a colon (:), followed by the name of the value label. If we read in this dataset and see what it contains

```
. webuse census9
(1980 Census data by state)

. describe
Contains data from http://www.stata-press.com/data/r8/census9.dta
  obs:            50                          1980 Census data by state
  vars:            5                          16 Jul 2002 18:29
  size:         1,550 (99.9% of memory free)

              storage  display    value
variable name   type   format     label    variable label

state          str14   %-14s                State
drate          float   %9.0g                Death Rate
pop            long    %12.0gc              Population
medage         float   %9.2f                Median age
region         byte    %-8.0g     cenreg    Census region

Sorted by:
```

we notice that variable `region` has values labeled by the `cenreg` value label. The correspondence between the underlying number and the value label is shown by

```
. label list
cenreg:
           1 NE
           2 N Cntrl
           3 South
           4 West
```

[R] **regress** uses this dataset to illustrate weighted regression. To obtain the regression of `drate` and `medage` restricted to the "South" region, you could type

```
. regress drate medage [aweight=pop] if region == 3
```

But, if you do not remember the underlying region number for "South", you could also obtain this regression by typing

```
. regress drate medage [aweight=pop] if region == "South":cenreg
(sum of wgt is    7.4734e+07)
```

Source	SS	df	MS		
Model	1072.30989	1	1072.30989		
Residual	550.163155	14	39.2973682		
Total	1622.47305	15	108.16487		

```
                                          Number of obs =      16
                                          F( 1,    14) =   27.29
                                          Prob > F      =  0.0001
                                          R-squared     =  0.6609
                                          Adj R-squared =  0.6367
                                          Root MSE      =  6.2688
```

drate	Coef.	Std. Err.	t	P>|t|	[95% Conf. Interval]
medage	3.905819	.7477109	5.22	0.000	2.302139 5.509499
_cons	-29.34031	22.33676	-1.31	0.210	-77.2479 18.56727

Typing the value label instead of the underlying number makes it unlikely that you will obtain an unintended result from entering the wrong region number. An added benefit of using the value label is that, when you later review your results, you will quickly see that the regression is for the "South" region, and you will not need to remember what region was assigned number 3.

See [U] **13.9 Label values** for further information about specifying value labels in expressions.

The Stata Journal (2004)
4, Number 4, pp. 488–489

Stata tip 15: Function graphs on the fly

Nicholas J. Cox
University of Durham, UK
n.j.cox@durham.ac.uk

[G] **graph twoway function** gives several examples of how function graphs may be drawn on the fly. The manual entry does not quite explain the full flexibility and versatility of the command. Here is a further advertisement on its behalf. To underline a key feature: you do not need to create variables following particular functions. The command handles all of that for you. We will look at two more examples.

A common simple need is to draw a circle. A trick with `twoway function` is to draw two half-circles, upper and lower, and combine them. If you are working in some scheme using color, you will usually also want to ensure that the two halves are shown in the same color. The `aspect()` option was explained in Cox (2004).

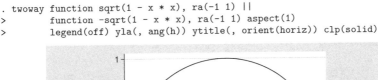

```
. twoway function sqrt(1 - x * x), ra(-1 1) ||
>          function -sqrt(1 - x * x), ra(-1 1) aspect(1)
>          legend(off) yla(, ang(h)) ytitle(, orient(horiz)) clp(solid)
```

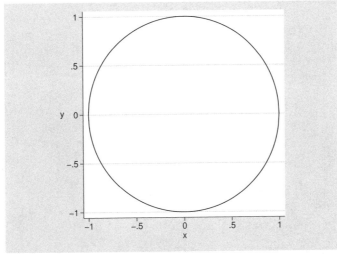

MacKay (2003, 316) asserts that, if we transform beta distributions of variables P between 0 and 1 to the corresponding densities over logit $P = \ln[P/(1 - P)]$, then we find always pleasant bell-shaped densities. In contrast, densities over P may have singularities at $P = 0$ and $P = 1$. This is the kind of textbook statement that should provoke some play with friendly statistical graphics software.

To explore MacKay's assertion, we need a standard result on changing variables (see, for example, Evans and Rosenthal 2004, theorems 2.6.2 and 2.6.3). Suppose that P is an absolutely continuous random variable with density function f_P, h is a function that is differentiable and monotone, and $X = h(P)$. The density function of X is then

$$f_X(x) = \frac{f_P\{h^{-1}(x)\}}{|h'\{h^{-1}(x)\}|}$$

In our case, $h(P) = \text{logit } P$, so that $h^{-1}(X) = \exp(X)/\{1+\exp(X)\}$ and $h'\{h^{-1}(X)\} = \{1 + \exp(X)\}^2/\exp(X)$. In Stata terms, beta densities transformed to the logit scale are the product of `betaden(p)` or `betaden(invlogit(x))` and `exp(x)/(1+exp(x))^2`. The latter term may be recognized as a logistic density function, which always has a bell shape.

An example pair of original and transformed distributions is given by the commands below. To explore further in parameter space, you need only vary the parameters from 0.5 and 0.5 (and, if desired, to vary the range).

```
. twoway function betaden(0.5,0.5,x), ytitle(density) xtitle(p)
. twoway function betaden(0.5,0.5,invlogit(x)) * (exp(x) / (1 + exp(x))^2),
> ra(-10 10) ytitle(density) xtitle(logit p)
```

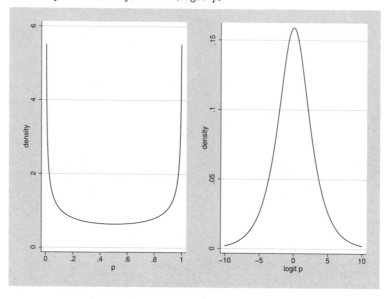

References

Cox, N. J. 2004. Stata tip 12: Tuning the plot region aspect ratio. *Stata Journal* 4: 357–358.

Evans, M. J., and J. S. Rosenthal. 2004. *Probability and Statistics: The Science of Uncertainty*. New York: Freeman.

MacKay, D. J. C. 2003. *Information Theory, Inference, and Learning Algorithms*. Cambridge: Cambridge University Press. Also available at http://www.inference.phy.cam.ac.uk/mackay/itprnn/book.html.

The Stata Journal (2005)
5, Number 1, p. 134

Stata tip 16: Using input to generate variables

Ulrich Kohler
Wissenschaftszentrum Berlin für Sozialforschung
kohler@wz-berlin.de

Sometimes using `generate` is an untidy and long-winded way to generate new variables, particularly if the variable you want to create is categorical and there are many different categories. Thus rather than using

```
. gen iso3166_2 = "AT" if country == "Austria"
. replace iso3166_2 = "BE" if country == "Belgium"
. replace iso3166_2 = "TR" if country == "Turkey"
```
and so on for say 28 countries
```
. gen iso3166_3 = "AUT" if country == "Austria"
. replace iso3166_3 = "BEL" if country == "Belgium"
. replace iso3166_3 = "TUR" if country == "Turkey"
```
and so on for say 28 countries
```
. gen gername = "Österreich" if country == "Austria"
. replace gername = "Belgien" if country == "Belgium"
. replace gername = "Türkei" if country == "Turkey"
```
and so on for say 28 countries

you can use `input` to produce a new dataset, `save` to a temporary file, and then `merge`:

```
. preserve
. clear
. input str15 country str2 iso3166_2 str3 iso3166_3 str15 gername
Austria AT AUT Österreich
Belgium BE BEL Belgien
Turkey  TK TUR Türkei
```
and so on
```
. end

. sort country
. tempfile foo
. save 'foo'
. restore
. sort country
. merge country using 'foo'
```

Among the benefits are less typing; a cleaner log file; in huge datasets, faster data processing; and arguably fewer errors.

See [R] **input**[1] for the finer points on `input`.

---

1. For Stata 9 and later, see [D] **input**.

The Stata Journal (2005)
5, Number 1, pp. 135–136

Stata tip 17: Filling in the gaps

Nicholas J. Cox
University of Durham, UK
n.j.cox@durham.ac.uk

The `fillin` command (see [R] **fillin**[1]) does precisely one thing: it fills in the gaps in a rectangular data structure. That is very well explained in the manual entry, but people who do not yet know the command often miss it, so here is one more plug. Suppose that you have a dataset of people and the choices they make, something like this:

```
id   choice
1    1
2    3
3    1
4    2
```

Now suppose that you wish to run a nested logit model using `nlogit` (see [R] **nlogit**). This command requires all choices, those made and those not made, to be explicit. With even 4 values of `id` and 3 values of `choice`, we need 12 observations so that each combination of variables exists once in the dataset; hence, 8 more are needed in this case. The solution is just

```
. fillin id choice
```

and a new variable, `_fillin`, is added to the dataset with values 1 if the observation was "filled in" and 0 otherwise. Thus `count if _fillin` tells you how many observations were added. You will often want to `replace` or `rename _fillin` to something appropriate:

```
. rename _fillin chosen
. replace chosen = 1 - chosen
```

If you do not `rename` or `drop _fillin`, it will get in the way of a subsequent `fillin`. Usually, the decision is clear-cut: Either `_fillin` has a natural interpretation, so you want to keep it, or a relative, under a different name; or `_fillin` was just a by-product, and you can get rid of it without distress.

Another common variant is to show zero counts or amounts explicitly. With a dataset of political donations for several years, we might want an observation showing that `amount` is zero for each pair of `donor` and `year` not matched by a donation. This typically leads to neater tables and graphs and may be needed for modeling: in particular, for panel models, the zeros must be present as observations. The main idea is the same, but the aftermath is different:

```
. fillin donor year
. replace amount = 0 if _fillin
```

---

1. For Stata 9 and later, see [D] **fillin**.

Naturally if we have more than one donation from various donors in various years, we might also want to `collapse` (or just possibly `contract`) the data, but that is the opposite kind of problem.

Yet another common variant is the creation of a grid for some purpose, perhaps before data entry, or before you draw a graph. You can be very lazy by typing

```
. clear
. set obs 20
. gen y = _n
. gen x = y
. fillin y x
```

which creates a 20×20 grid. This is good, but sometimes you want something different; see functions `fill()` and `seq()` in [R] **egen**.[2]

The messiest `fillin` problems are when some of the categories you want are not present in the dataset at all. If you know a person is not one of the values of **donor**, no amount of filling in will add a set of zeros for that person. One strategy here is to add pseudo-observations so that every category occurs at least once and then to `fillin` in terms of that larger dataset. This is just a variation on the technique for creating a grid out of nothing.

As far as you can see from what is here, `fillin` just does things in place, so you need not worry about file manipulation. This is an illusion, as underneath the surface, `fillin` is firing up `cross` (see [R] **cross**[3]), which does the work using files. Thus `cross` is more fundamental. A forthcoming tip will say more.

---

2. For Stata 9 and later, see [D] **egen**.
3. For Stata 9 and later, see [D] **cross**.

The Stata Journal (2005)
5, Number 1, pp. 137–138

Stata tip 18: Making keys functional

Shannon Driver
StataCorp
s.driver@stata.com

Did you know that you can create custom definitions for your *F*-keys in Stata?

F-key definitions are created via global macros. On startup, Stata sets the *F*-key defaults to

F-key	definition
F1	`help`
F2	`#review;`
F3	`describe;`
F7	`save`
F8	`use`

You can redefine these keys if you wish.

When a definition ends with a semicolon (;), Stata will automatically execute that command as if you typed it and pressed the Enter key; otherwise, the command is immediately entered into the command line as if you had typed it. Stata then waits for you to press the Enter key. This allows you to modify the command before it is executed.

For example, to define the *F4* key to execute the `list` command, you would type

```
. global F4 "list;"
```

The "F4" here is actually a capital `F` followed by the number 4.

The best place to create these definitions is in an ASCII text file called `profile.do`. Every time Stata is launched, it looks for `profile.do` and, if it finds it, executes all of the commands it contains. For more information, type `help profile`.

Let's say that you want to create a definition for *F4* to open a window showing the contents of a particular directory. You could do this on Windows by typing

```
. global F4 '"winexec explorer C:\data;"'
```

On a Macintosh, you could type

```
. global F4 '"!open /Applications/Stata8/Stata;"'
```

You can also create *F*-key definitions to launch your favorite text editor.

```
. global F5 '"winexec notepad;"'
```

Yet another application is programming the ' and ' keys, which Stata uses to delimit local macros. Many keyboards do not have the left- or open-quote character of this pair, so an alternative is to define an *F*-key to be that key. For symmetry, you might want another *F*-key to be the right- or close-quote character. But how do you define a replacement for a key if you do not have that key in the first place? One answer lies in Stata's `char()` function:

```
. global F4 = char(96)
. global F5 = char(180)
```

You may want to make a note that *F10* is reserved internally by Windows, so you cannot program this key. Also, not all Macintosh keyboards have *F*-keys.

For more information on this topic, please see [U] **10.2 F-keys**.

The Stata Journal (2005)
5, Number 2, p. 279

Stata tip 19: A way to leaner, faster graphs

Patrick Royston
MRC Clinical Trials Unit
p.royston@ctu.mrc.ac.uk

If you have many variables, consider doing a `preserve` of the data and `dropping` several of them before drawing a graph. This greatly speeds up production.

Take plotting fitted values from a model as an example. If there are many tied observations at each value of the predictor and therefore many replicates of the fitted values, the size of the graph file can be large, also making the plotting time large. A construction like this can save resources:

```
. preserve
. bysort x: drop if _n > 1
. line f1 f2 f3 x, sort clp(l - _) saving(graph, replace)
. restore
```

Here is another real example: with 15,156 variables and 50 observations, I wanted a `dotplot` of variable v15155 by v15156. The time taken with all data present was 10.66 seconds, but with `preserve` and all irrelevant variables `dropped`, it was 0.69 seconds.

The Stata Journal (2005)
5, Number 2, pp. 280–281

Stata tip 20: Generating histogram bin variables

David A. Harrison
ICNARC, London, UK
david.harrison@icnarc.org

Did you know about `twoway__histogram_gen`? (Note the two underscores in the first gap and only one in the second.) This command is used by `histogram` to generate the variables that are plotted. It is undocumented in the manuals but explained in the online help. The command can be used directly to save these variables, enabling more complex manipulation of histograms and production of other graphs or tables.

Consider the S&P 500 historical data that are used as an example for [R] **histogram**:

```
. use http://www.stata-press.com/data/r9/sp500
(S&P 500)
. histogram volume, percent start(4000) width(1000)
(bin=20, start=4000, width=1000)
  (output omitted)
```

To display only the central part of this histogram from 8,000 to 16,000, we could use `if`, but this will change the height of the bars, as data outside the range 8,000 to 16,000 will be ignored completely. To restrict the range without altering the bars, we use `twoway__histogram_gen` to save the histogram and only plot the section of interest:

```
. twoway__histogram_gen volume, percent start(4000) width(1000) gen(h x)
. twoway bar h x if inrange(x,8000,16000), barwidth(1000) bstyle(histogram)
```

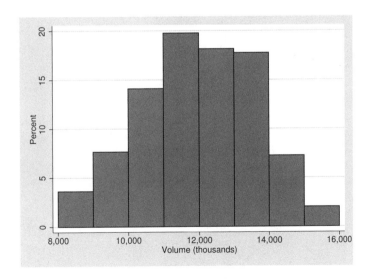

The `start()` and `width()` options above specified cutpoints that included 8,000 and 16,000. We could, alternatively, use the default cutpoints:

```
. twoway__histogram_gen volume if inrange(volume,8000,16000), display
(bin=14, start=8117, width=525.08571)
. local m = r(start)
. local w = r(width)
. summarize volume, meanonly
. local s = 'm' - 'w' * ceil(('m' - r(min))/'w')
. twoway__histogram_gen volume, percent start('s') width('w') gen(h x, replace)
. twoway bar h x if inrange(x,8000,16000), barwidth('w') bstyle(histogram)
  (output omitted)
```

Other uses of `twoway__histogram_gen` include the following:

- Overlaying or mirroring two histograms

```
. use http://www.stata-press.com/data/r9/bplong, clear
(fictional blood-pressure data)
. twoway__histogram_gen bp if sex == 0, frac start(125) w(5) gen(h1 x1)
. twoway__histogram_gen bp if sex == 1, frac start(125) w(5) gen(h2 x2)
. twoway (bar h1 x1, barw(5) bc(gs11))
> (bar h2 x2, barw(5) blc(black) bfc(none)),
> legend(order(1 "Male" 2 "Female"))
  (output omitted)
. qui replace h2 = -h2
. twoway (bar h1 x1, barw(5)) (bar h2 x2, barw(5)),
> yla(-.2 ".2" -.1 ".1" 0 .1 .2) legend(order(1 "Male" 2 "Female"))
  (output omitted)
```

- Changing the scale, for example, to plot density on a square-root scale

```
. twoway__histogram_gen bp, start(125) width(5) gen(h x)
. qui gen hsqrt = sqrt(h)
. twoway bar hsqrt x, barw(5) bstyle(histogram) ytitle(Density)
> ylabel(0 .05 ".0025" .1 ".01" .15 ".0225" .2 ".04")
  (output omitted)
```

- Plotting the differences between observed and expected frequencies

```
. twoway__histogram_gen bp, freq start(125) w(5) gen(h x, replace)
. qui summarize bp
. qui gen diff = h - r(N) * (norm((x + 2.5 - r(mean))/r(sd)) -
> norm((x - 2.5 - r(mean))/r(sd)))
. twoway bar diff x, barw(5) yti("Observed - expected frequency")
  (output omitted)
```

There are also two similar commands: `twoway__function_gen` to generate functions and `twoway__kdensity_gen` to generate kernel densities.

The Stata Journal (2005)
5, Number 2, pp. 282–284

Stata tip 21: The arrows of outrageous fortune

Nicholas J. Cox
Durham University
n.j.cox@durham.ac.uk

Stata 9 introduces a clutch of new plottypes for `graph twoway` for paired-coordinate data. These are defined by four variables, two specifying starting coordinates and the other two specifying ending coordinates. Here we look at some of the possibilities opened up by [G] **graph twoway pcarrow** for graphing changes over time. Arrows are readily understood by novices as well as experts as indicating, in this case, the flow from the past towards the present.

Let us begin with one of the classic time-series datasets. The number of lynx trapped in an area of Canada provides an excellent example of cyclic boom-and-bust population dynamics. Trappings are optimistically assumed to be proportional to the unknown population size.

The dataset has already been `tsset` (see [TS] **tsset**).

```
. use http://www.stata-press.com/data/r9/lynx2.dta, clear
(TIMESLAB: Canadian lynx)

. tsline lynx
  (output omitted)
```

Ecologists and other statistically minded people find it natural to think about populations on a logarithmic scale: population growth is after all inherently multiplicative. Logarithms to base 10 are convenient for graphing.

(Continued on next page)

```
. gen loglynx = log10(lynx)

. twoway pcarrow loglynx L.loglynx F.loglynx loglynx,
> xla(2 "100" `=log10(200)' "200" `=log10(500)' "500" 3 "1000" `=log10(2000)'
> "2000" `=log10(5000)' "5000")
> yla(2 "100" `=log10(200)' "200" `=log10(500)' "500" 3 "1000" `=log10(2000)'
> "2000" `=log10(5000)' "5000")
> ytitle(this year) xtitle(previous year) subtitle(Number of lynx trapped)
```

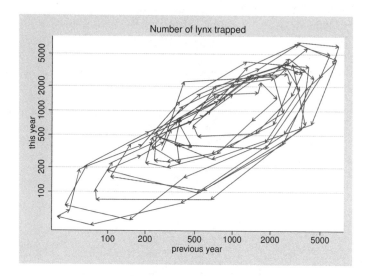

Thinking as it were autoregressively, we can plot this year's population versus the previous year's and join data points with arrows end to end. Each data point other than the first and last is the end of one arrow pointing from `L.loglynx` to `loglynx` and the beginning of another pointing from `loglynx` to `F.loglynx`. This dual role and the ability to use time-series operators such as `L.` and `F.` on the fly in graphics commands yield the command syntax just given. Plotted as a trajectory in this space, population cycles are revealed clearly as asymmetric. Depending on your background, you may see this as an example of hysteresis, or whatever else it is called in your tribal jargon.

Another basic comparison compares values for some outcome of interest at two dates. For this next example, we use life expectancy data for 1970 and 2003 from the UNICEF report, *The State of the World's Children 2005*, taken from the web site *http://www.unicef.org* accessed on May 12, 2005. A manageable graph focuses on those countries for which life expectancy was under 50 years in 2003. A `count` on the dataset thus entered shows that there are 33 such countries.

We borrow some ideas from displays possible with `graph dot` (see [G] **graph dot**). Arrows connecting pairs of variables are not supported by `graph dot`. However, as is common with Stata's graphics, whatever is difficult with `graph dot`, `graph bar`, or `graph hbar` is often straightforward with `graph twoway`, modulo some persistence.

A natural sort order for the graph is that of life expectancy in 2003. A nuance to make the graph tidier is to break ties according to life expectancy in 1970. Life expectancy is customarily, and sensibly, reported in integer years, so ties are common. One axis for the graph is then just the observation number given the sort order, except that we will want to name the countries concerned on the graph. For names that might be fairly long, we prefer horizontal alignment and thus a vertical axis. The names are best assigned to value labels. Looping over observations is one way to define those. The online help on `forvalues` and `macros` explains any trickery with the loop that may be unfamiliar to you; also see [P] **forvalues** and [P] **macro**.

```
. gsort lifeexp03 - lifeexp70
. gen order = _n
. forval i = 1/33 {
  2.        label def order 'i' "'=country['i']'", modify
  3. }
. label val order order
```

The main part of the graph is then obtained by a call to `twoway pcarrow`. The arrowhead denotes life expectancy in 2003. Optionally, although not essentially, we overlay a scatter plot showing the 1970 values.

```
. twoway pcarrow order lifeexp70 order lifeexp03 if lifeexp03 < 50
> || scatter order lifeexp70 if lifeexp03 < 50, ms(oh)
> yla(1/33, ang(h) notick valuelabel labsize(*0.75)) yti("") legend(off)
> barbsize(2) xtitle("Life expectancy in years, 1970 and 2003") aspect(1)
```

Apart from Afghanistan, all the countries shown are in Africa. Some show considerable improvements over this period, but in about as many, life expectancy has fallen dramatically. Readers can add their own somber commentary in terms of war, political instability, famine, and disease, particularly AIDS.

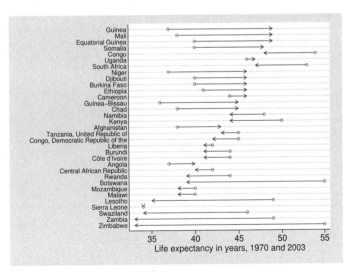

The Stata Journal (2005)
5, Number 3, pp. 465–466

Stata tip 22: Variable name abbreviation

Philip Ryan
University of Adelaide
philip.ryan@adelaide.edu.au

Stata allows users to abbreviate any variable name to the shortest string of characters that uniquely identifies it, given the data currently loaded in memory (see [U] **11.2.3 Variable-name abbreviation**). Stata also offers three wildcard characters, *, ~, and ? (see [U] **11.4.1 Lists of existing variables**), so users have substantial flexibility in how variables may be referenced.

Stata also allows users to control the values of many of its system parameters using the `set` command (see [R] **set**). One of these parameters is `varabbrev`, which may be toggled on, allowing variable names to be abbreviated, or off, requiring the user to spell out entire variable names.

The default is to allow abbreviations. But this convenience feature can bite. Suppose that in a program we wish to confirm the existence of a variable and that variable does not in fact exist:

```
. clear
. set varabbrev on
. set obs 10
(obs was 0, now 10)
. generate byte myvar7 = 1
. confirm variable myvar
```

There is no error message here because `myvar` is an allowed abbreviation for `myvar7`. A bigger deal is that as `myvar7` exists and not `myvar`, typing `drop myvar` would drop `myvar7`, which may or may not have been our intention.

But what if we had wanted to confirm explicitly the existence of variable `myvar`? There are two ways to do this:

1. Specify the `confirm` command with the `exact` option (see [P] **confirm**):

   ```
   . confirm variable myvar, exact
   variable myvar not found
   r(111);
   ```

2. Toggle variable abbreviation off:

   ```
   . set varabbrev off
   . confirm variable myvar
   variable myvar not found
   r(111);
   ```

Note that the status of `varabbrev` does not affect the display of variable names. For example,

```
. sysuse auto, clear
(1978 Automobile Data)

. set varabbrev off

. rename weight this_is_a_very_long_varname

. regress price turn length this_is_a_very_long_varname
   (output omitted)
```

price	Coef.	Std. Err.	t	P>|t|	[95% Conf. Interval]	
turn	-318.2055	127.1241	-2.50	0.015	-571.7465	-64.66452
length	-66.17856	39.87361	-1.66	0.101	-145.704	13.34684
this_is_a_~e	5.382135	1.116756	4.82	0.000	3.154834	7.609435
_cons	14967.64	4541.836	3.30	0.002	5909.228	24026.04

In this display, Stata has abbreviated the long variable name, despite the current value of `varabbrev`.

Note that the `list` command has its own option to allow the user partial control of the display; see [D] **list**. As we `set varabbrev off`, we must specify only unabbreviated variable names in a `list` command, but we can override Stata's default abbreviation in the display using the `abbreviate()` option:

```
. list make turn this_is_a_very_long_varname in 1/4, abb(21)
```

	make	turn	this_is_a_very_long_~e
1.	AMC Concord	40	2,930
2.	AMC Pacer	40	3,350
3.	AMC Spirit	35	2,640
4.	Buick Century	40	3,250

The default value for `abbreviate()` is 8, so that otherwise the variable name would have been displayed as `this_i~e`.

The Stata Journal (2005)
5, Number 3, pp. 467–468

Stata tip 23: Regaining control over axis ranges

Nicholas J. G. Winter
Cornell University
nw53@cornell.edu

Beginning with version 8, Stata will often widen the range of a graph axis beyond the range of the data. Convincing Stata to narrow the range can be difficult unless you understand the cause of the problem.

Using the trusty `auto` dataset, consider the graph produced by this command:

```
. sysuse auto, clear
(1978 Automobile Data)

. twoway scatter mpg price
```

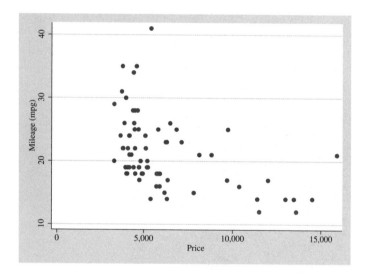

Although price ranges from \$3,291 to \$15,906 in the data, the lower end of the x-axis in this graph extends to zero, leaving blank space on the left-hand side. If we do not like this space, the solution would seem to lie with the `range()` suboption of the `xscale()` option. Thus we might expect Stata to range the x-axis using the minimum and maximum of the data, given the following command:

```
. twoway scatter mpg price, xscale(range(3291 15906))
```

However, this produces the same graph: the axis still includes zero. It seems that Stata is ignoring `range()`, although it does not do that when the range is *increased*, rather than decreased. Consider, for example, this command, which expands the x-axis to run from 0 through 30,000:

```
. twoway scatter mpg price, xscale(range(0 30000))
```

The issue is that the range displayed for an axis depends on the interaction between two sets of options (or their defaults): those that control the axis range explicitly, and those that *label* the axis. The range can be expanded either by explicitly specifying a longer axis (e.g., with xscale(range(a b))) or by labeling values outside the range of the data.

To determine the range of an axis, Stata begins with the minimum and maximum of the data. Then it will widen (but never narrow) the axis range as instructed by range(). Finally, it will widen the axis if necessary to accommodate any axis labels.

By default, twoway labels the axes with "about" five ticks, the equivalent of specifying xlabel(#5). In this case, Stata chooses four labels, one of which is zero, and then expands the *x*-axis accordingly. In other words, if we specify xscale()—but do not specify xlabel()—we are in effect saying to Stata "and please use the default xlabel() for this graph". This default may widen the axis range.

Therefore, to get a narrower *x*-axis, we must specify a narrower set of axis labels. For example, to label just the minimum and maximum, we could specify

```
. scatter mpg price, xlabel(minmax)
```

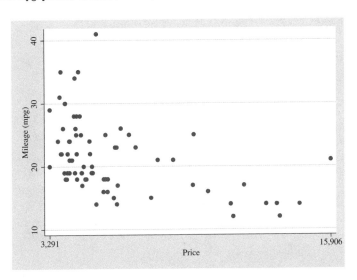

Of course, we could specify any other set of points to label, for example,

```
. scatter mpg price, xlabel(5000[1000]15000)
```

This issue did not appear until Stata version 8. Prior versions defaulted to labeling the minimum and the maximum of the data only. This would be equivalent to including the options xlabel(minmax) and ylabel(minmax) in Stata 8 or later.

The Stata Journal (2005)
5, Number 3, p. 469

Stata tip 24: Axis labels on two or more levels

Nicholas J. Cox
Durham University
n.j.cox@durham.ac.uk

Text shown as graph axis labels is by default shown on one level. For example, a label `Foreign cars` would be shown just like that. Sometimes you want the text of a label to be shown on two or even more levels, as one way of reducing crowding or even overprinting of text; thus you might want `Foreign` written above `cars`. Other ways of fighting crowding include varying the size or angle at which text is printed (see [G] *axis_label_options* for details), or in some cases reconsidering which variable should go on which axis.

To specify multiple levels, the text to go on each level should appear within double quotes " ", and the whole text label should appear within compound double quotes ` " " `. For more explanation of the latter, see [U] **18.3.5 Double quotes**. That way, Stata's parser has a clear idea of parts and wholes.

Here are some examples:

```
. sysuse auto
. dotplot mpg, over(foreign)
> xlabel(0 `" "Domestic" "cars" "' 1 `" "Foreign" "cars" "') xtitle("")
. graph box mpg,
> over(foreign, relabel(1 `" "Domestic" "cars" "'  2 `" "Foreign" "cars" "'))
. graph hbar (mean) mpg,
> over(foreign, relabel(1 `" "Domestic" "cars" "'  2 `" "Foreign" "cars" "'))
```

Note the subtle difference between these examples. `dotplot` is really a wrapper for `twoway` and, as is characteristic of `twoway` graphs, it takes its variables literally so that the values of `foreign` are indeed treated as 0 and 1. On the other hand, graphs with so-called categorical axes (`graph bar`, `graph hbar`, `graph box`, `graph hbox`, and `graph dot`) consider the categories shown to be 1, 2, and so forth, regardless of the precise numeric or string values of the variables concerned. The numbers increase from left to right or from top to bottom, as the case may be. Thus matrix users will feel at home with this convention.

The Stata Journal (2005)
5, Number 4, pp. 602–603

Stata tip 25: Sequence index plots

Ulrich Kohler and Christian Brzinsky-Fay
Wissenschaftszentrum Berlin
kohler@wz-berlin.de, brzinsky-fay@wz-berlin.de

Sequence index plots of longitudinal or panel data use stacked bars or line segments to show how individuals move between a set of conditions or states over time. Changes of state are shown by changes of color. The term *sequence index plot* was proposed by Brüderl and Scherer (2004). See Scherer (2001) for an application.

It is possible to draw sequence index plots with Stata by using the `twoway` plottype `rbar`. Starting from data in survival-time form (see `help st`), you simply overlay separate range-bar plots for each state.

For example, suppose that you have data on times for entering and leaving various states of employment:

```
. list in 1/10
```

	id	type	begin	end
1.	1	employed	1	13
2.	1	apprenticeship	13	20
3.	1	unemployed	20	23
4.	1	employed	23	25
5.	1	unemployed	25	26
6.	1	employed	26	43
7.	1	unemployed	43	50
8.	1	employed	50	60
9.	2	employed	1	13
10.	2	apprenticeship	13	21

First, **separate** the start and end dates for the different states:

```
. separate begin, by(type)
. separate end, by(type)
```

Then plot overlaid range bars for each state:

```
. graph twoway
>    (rbar begin1 end1 id, horizontal)
>    (rbar begin2 end2 id, horizontal)
>    (rbar begin3 end3 id, horizontal)
>    (rbar begin4 end4 id, horizontal)
>    (rbar begin5 end5 id, horizontal)
>    , legend(order(1 "education" 2 "apprenticeship"
>          3 "employment" 4 "unemployment" 5 "inactivity")
>          cols(1) pos(2) symxsize(5))
>      xtitle("months") yla(, angle(h)) yscale(reverse)
```

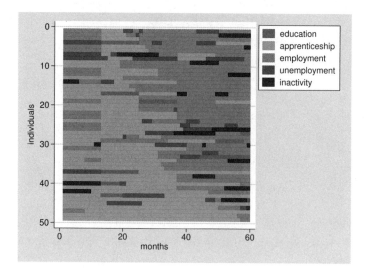

It is common to put personal identifiers on the *y*-axis, using the option `horizontal`, and put time on the *x*-axis.

In practice, with many individuals in a large panel, the bars become thinner lines. In such cases, you could use the plottype `rspike` instead. Note also that you can make room for more individuals by tuning the aspect ratio of the graph (see Cox 2004). There is no upper limit to how many individuals are shown, although as the number increases, the resulting graph may become too difficult to interpret. The readability, however, largely depends on how far similar individuals are grouped together. The sort order should therefore be some criterion of similarity between sequences.

To fine tune the graph, use any option allowed with `graph twoway`; type `help twoway_options`. Our example provides some simple illustrations. `legend()` changes the contents and placement of the legend. `xtitle()` defines the title along the *x*-axis. `ylabel()` is used to display the *y*-axis labels horizontally, instead of vertically. `yscale(reverse)` reverses the scale of the *y*-axis so that the first individual is plotted at the very top of the graph.

References

Brüderl, J., and S. Scherer. 2004. Methoden zur Analyse von Sequenzdaten. *Kölner Zeitschrift für Soziologie und Sozialpsychologie* Sonderheft 44: Methoden der Sozialforschung: 330–347.

Cox, N. J. 2004. Stata tip 12: Tuning the plot region aspect ratio. *Stata Journal* 4: 357–358.

Scherer, S. 2001. Early career patterns: A comparison of Great Britain and West Germany. *European Sociological Review* 17: 119–144.

The Stata Journal (2005)
5, Number 4, p. 604

Stata tip 26: Maximizing compatibility between Macintosh and Windows

Michael S. Hanson
Wesleyan University
mshanson@wesleyan.edu

A questioner on Statalist asked whether there are problems using Stata in a joint project on different operating systems. My short answer is "No". Underlying this is StataCorp's work to ensure that its official filetypes (`.dta`, `.gph`, `.ado`, `.do`, `.dct`, etc.) are completely compatible across all the operating systems it supports and that Stata can always read those files even from older versions.

My slightly longer answer is "Not with a few basic precautions", but much depends on how the collaboration occurs. Here is some advice for easier joint use of Stata across platforms (and indeed on the same platform).

Standardize Statas. Ideally, everyone should use the same version of Stata, with the same additional `.ado` files installed.

Avoid absolute file paths. (This strikes me as generally a good idea anyway, as it leads to greater portability.) If the project is sufficiently complex, create an identical subdirectory structure below some common project root directory on every file system and only use path references relative to this root.

Use forward slashes in file paths. Note that Stata understands the forward slash (/) as separating directory levels on all platforms, even Windows. So use that instead of the backward slash (\) for paths in all `.do` files.

Watch end-of-line delimiters. Text files have different line endings on Macintosh, Windows, and Unix systems. So long as users on different platforms are using sufficiently versatile text editors, it should be straightforward to read both input files (e.g., `.do` files) and output files (e.g., `.log` files) regardless of the line endings used—and, if necessary, to convert to the desired one. Note that how the files will be shared—common server, (S)FTP, email—may have implications for this issue.

Use Encapsulated PostScript. Save graphics in EPS format. (See the help or manual entry for `graph export`.) While the Macintosh can natively generate graphics in PDF format, the PC cannot (without jumping through some hoops and purchasing Acrobat, that is). WMF, EMF, and PICT do not translate well across platforms, and while you could use PNG, nonvector formats do not scale well. (As they are at a fixed size, when enlarged, they still contain only the lower-density information of the original, and when they are reduced, information must be lost to reduce their size.)

Use other open-standards file formats. More generally, collaboration will be easier if everyone uses open-standards file formats (e.g., plain text, PNG, TeX, or LaTeX, etc.)—instead of those tied to proprietary software.

The Stata Journal (2005)
5, Number 4, pp. 605–607

Stata tip 27: Classifying data points on scatter plots

Nicholas J. Cox
Durham University
n.j.cox@durham.ac.uk

When you have scatter plots of counted or measured variables, you may often wish to classify data points according to the values of a further categorical variable. There are several ways to do this. Here we focus on the use of `separate`, gray-scale gradation, and text characters as class symbols. If different categories really do plot as distinct clusters, it should not matter too much how you show them, but knowing some Stata tricks should also help.

One starting point is that differing markers may be used on the plot whenever there are several variables plotted on the y-axis. With the `auto.dta` dataset, you can imagine

```
. sysuse auto
. gen mpg0 = mpg if foreign == 0
. gen mpg1 = mpg if foreign == 1
. scatter mpg? weight
```

Note the use of the wildcard `mpg?`, which picks up any variable names that have `mpg` followed by just one other character. Once the two variables `mpg0` and `mpg1` have been generated, different markers are automatic. This process still raises two questions. To get an acceptable graph, we need self-explanatory variable labels or at least self-explanatory text in the graph legend. Moreover, two categories are easy enough, but do we have to do this for each of say 5, 7, or 9 categories?

In fact, it would have been better to type

```
. separate mpg, by(foreign) veryshortlabel
. scatter mpg? weight
```

The command `separate` (see [D] **separate**) generates all the variables we need in one command and has a stab at labeling them intelligibly. In this case, we use the (undocumented) `veryshortlabel` option, which was implemented with graphics especially in mind. You may prefer the results of the documented `shortlabel` option. Note that the `by()` option can take true-or-false conditions, such as `price < 6000`, as well as categorical variables.

If your categorical variable consists of qualitatively different categories, you are likely to want to use qualitatively different symbols. Alternatively, if that variable is ordered or graded, the coding you use should also be ordered. One possibility is to use symbols colored in a sequence of gray scales.

Some data on landforms illustrate the point: Ian S. Evans kindly supplied measurements of 260 cirques in Wales, armchair-shaped hollows formerly occupied by small glaciers. Length tends to increase with width, approximately as a power function, but qualitative aspects of form, particularly how closely they approach a classic, well-developed shape, are also coded in a grade variable.

```
. separate length, by(grade) veryshortlabel
. scatter length? width, xsc(log) ysc(log) ms(0 ..)
> mcolor(gs1 gs4 gs7 gs10 gs13) mlcolor(black ..) msize(*1.5 ..)
> yti("': variable label length'") yla(200 500 1000 2000, ang(h))
> xla(200 500 1000 2000) legend(pos(11) ring(0) col(1))
```

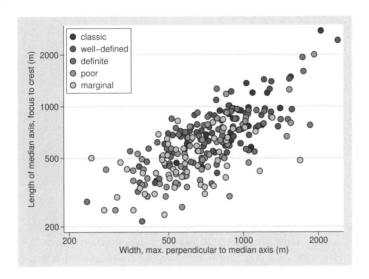

Figure 1 shows length versus width, subdivided by grade. Some practical details deserve emphasis. Gray scales near 16 (white) may be difficult to spot against a light background, including any printed page. Therefore, a dark outline color is recommended. Bigger symbols than the default are needed to do the coloring justice, but as a consequence, this approach is less likely to be useful with thousands of data points. A by() option showing different categories separately might work better. With the coding here, it so happens that the darkest category is plotted first and is thus liable to be overplotted by lighter categories wherever data points are dense. Some experimentation with the opposite order of plotting might be a good idea to see which works better.

An alternative that sometimes works nicely is to use ordinary text characters as different markers. One clean style is to suppress the marker symbols completely, using instead the contents of a str1 variable as marker labels. Whittaker (1975, 224) gave data on net primary productivity and biomass density for various ecosystem types. Figure 2 shows the subdivision.

```
. scatter npp bd, xsc(log) ysc(log) ms(i) mlabpos(0) mlabsize(*1.4)
> mla(c) yla(3000 1000 300 100 30 10 3, nogrid ang(h))
> xla(0.01 "0.01" 0.1 "0.1" 1 10 100)
> legend(on ring(0) pos(5) order( - "m marine" - "w wet" - "c cultivated" -
> "g grassland" - "f forest" - "b bare"))
```

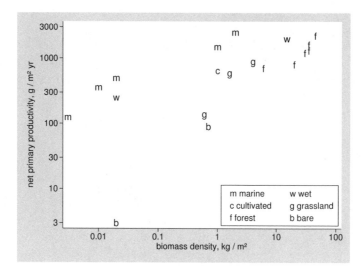

With three or four orders of magnitude variation in each variable, log scales are advisable. On those scales, there is a broad correlation whereby more biomass means higher productivity, but also considerable variation, much of which can be rationalized in terms of very different cover types. For the same biomass density, marine and other wet ecosystems have higher productivity than land ecosystems.

On the Stata side, remember `mlabpos(0)` and note that the `legend` must be set `on` explicitly. For different purposes, or for different tastes, what is here given as the legend might go better as text in a caption in a printed report. Behind the practice here lies general advice that lowercase letters, such as `abc`, work better than uppercase, such as `ABC`, as they are easier to distinguish from each other, and they are less likely to impart an synaesthetic sense in readers that the graph designer is shouting at them.

Reference

Whittaker, R. H. 1975. *Communities and Ecosystems*. New York: Macmillan.

The Stata Journal (2006)
6, Number 1, pp. 144–146

Stata tip 28: Precise control of dataset sort order

L. Philip Schumm
Department of Health Studies
University of Chicago
Chicago, IL
pschumm@uchicago.edu

The observations in a Stata dataset are ordered, so that they may be referred to by their position (e.g., `in 42/48`) and that individual values of a variable may be referred to with subscripts (e.g., `mpg[42]`). This order can be changed by using the `sort` command (see [D] `sort`). Developing a full appreciation of what is possible using `sort` together with the `by:` prefix, the underscore built-ins `_n` and `_N`, and subscripting is a major step toward Stata enlightenment (e.g., see Cox [2002]).

One source of surprise for many users arises when sorting by one or more variables which, when taken together, do not uniquely determine the order of observations. In this case, the resulting order within any group of observations having the same value(s) of those variables is effectively random because `sort` uses an *unstable* sort algorithm. Users who desire a *stable* sort—in which the previous ordering of observations within tied values of the sort variables is maintained—should specify the `stable` option. However, this option will slow `sort` down and, more importantly, can hide problems in your code.

You are likely to discover this issue when coding an operation dependent on the order of the data that gives different results from one run to another. Consider the following dataset consisting of mothers and their children:

```
. list, sepby(family)
```

	family	name	child
1.	2	Harriet	0
2.	2	Lewis	1
3.	1	Sylvia	0
4.	1	Jenny	1
5.	3	Kim	0
6.	3	Peter	1
7.	3	Kim	1

Individuals are grouped by family, the mother always appearing first. Suppose that we want to construct a unique within-family identifier, such that all mothers have the same value. This is a straightforward application of `by:`, but first the data must be sorted by family:

```
. sort family
. by family: generate individual = _n
```

```
. table child individual
```

| | individual | | |
child	1	2	3
0	2	1	
1	1	2	1

Unfortunately, the result is not as desired: one mother was assigned the value 2. In fact, following the call to `sort`, the order of observations within families—and hence the assignment of identifiers—was random. If we had instead sorted by family *and* child, each mother would have appeared first and would have been assigned a value of 1 (assuming that each family has exactly one mother—a key assumption that should always be checked). Yet even this solution would still be deficient: if a family has multiple children, their identifiers would be random and irreproducible. Only if we sort by family, child, *and* name would we have an adequate solution.

If we had used instead

```
. sort family, stable
```

we would also have obtained the desired result. So why does `sort` by default perform an unstable sort? Apart from better performance, the answer (emphasized by William Gould on Statalist) is that using the `stable` option not only fails to address the problem; it also reduces the chance of discovering it. Our error was to perform a calculation dependent on the sort order of the data without establishing that order beforehand. Using `stable` would have temporarily masked the error. However, had the sort order of the input dataset changed, we would have been in trouble.

How can you avoid such problems? First, train yourself to recognize when a calculation depends on the sort order of the data. Most instances in which you are using _n and _N or subscripting (either alone or with `by`) are easy to recognize. However, instances in which you are using a function that depends on the order of the data (e.g., `sum()` or `group()`) can be more subtle (Gould 2000).

Second, ensure that the order of the data is fully specified. This check became much easier in Stata 8 with the introduction of the `isid` command ([D] **isid**), which checks whether one or more variables uniquely identify the observations and returns an error if they do not. The command also has a `sort` option, which sorts the dataset in order of the specified variable(s). This option lets us replace our original `sort` command with

```
. isid family child name, sort
```

which, since it runs without error, confirms that we have specified the order fully. Had we used only `family`, or `family` and `child`, `isid` would have returned an error, immediately alerting us to the problem.

References

Cox, N. J. 2002. Speaking Stata: How to move step by: step. *Stata Journal* 2: 86–102.

Gould, W. 2000. FAQ: Sorting on categorical variables.
http://www.stata.com/support/faqs/lang/sort.html.

The Stata Journal (2006)
6, Number 1, pp. 146–148

Stata tip 29: For all times and all places

Charles H. Franklin
Department of Political Science
University of Wisconsin–Madison
Madison, WI
chfrankl@wisc.edu

According to the *Data Management Reference Manual*, the `cross` command is "rarely used"; see [D] **cross**. This comment understates the command's usefulness. For example, the `fillin` command uses `cross` (Cox 2005). Here is one further circumstance in which it proves extremely useful, allowing a simple solution to an otherwise awkward problem.

In pooled time-series cross-sectional data, we require that some number of units (geographic locations, patients, television markets) be observed over some period (daily from March to November, say). We thus need a data structure in which each unit is represented at each time point. If the data come in this complete form, then no problem arises. But when aggregating from lower-level observations, some dates, and possibly some units, are often missing. This missingness could be because no measurement was taken or because an event that is being counted simply did not occur on that date and so no record or observation was generated. In the aggregated Stata data file, no observation will appear for these dates or units. Inserting observations for the missing dates or units is awkward, but the `cross` command, followed by `merge`, makes the solution simple.

To illustrate with a real example: in the Wisconsin Advertising Project, we have coded 1.06 million political advertisements broadcast during the 2004 U.S. presidential campaign, using data provided by Nielsen Monitor-Plus. These ads are distributed across 210 media markets. Each time an ad is broadcast, it generates an observation in our dataset. The data are then aggregated to the media market to produce a daily count of the total advertising in each market. Such aggregation is simple in Stata. Variables `repad` and `demad` are coded 1 if the ad supported the Republican or Democratic candidate, respectively, and 0 otherwise. The sum is thus simply the count of the number of ads supporting each candidate.

```
clear
use allads
sort market date
collapse (sum) repad demad, by(market date) fast
save marketcounts, replace
```

This do-file produced no observation if no ads ran in a market on a particular date, which is common in these data. We want a dataset that includes every date for each of the 210 markets, with a value of 0 if no ad ran in a market on a date.

We can use `cross` to create a dataset that has one observation for each market for each of the 245 days included in our study. The file `dmacodelist.dta` contains one observation for each of the 210 markets: `dma` stands for "designated market area", Nielsen's term for television markets. First, we create a Stata dataset with 245 observations, one for each day of our study (March 3–November 2). Then we convert this information to a Stata date.

```
clear
set obs 245
gen date = _n + mdy(03,02,2004)
format date %d
```

Now use `cross` to generate the dataset with all dates for all markets:

```
cross using dmacodelist
sort market date
save alldates, replace
```

The file `alldates.dta` contains one observation for each market and for each date. The last step is to merge the aggregated `marketcount.dta` dataset with `alldates.dta` and replace missing values with zeros.

```
clear
use marketcounts
sort market date
merge market date using alldates
assert _merge != 1
replace demad = 0 if demad == .
replace repad = 0 if repad == .
```

The merge should produce no values of `_merge` that are 1, meaning observations found only in `marketcounts`, so the `assert` command checks this: the do-file will stop if the assertion is false (see Gould 2003 on `assert`). The `repads` and `demads` will be missing in the merged data only if no ad was broadcast, so replacing missing values for these variables with zeros will result in the desired dataset.

Thus the `cross` command offers an efficient solution to this type of problem. Those who often aggregate low-level data to create time-series cross-sectional structures will find this command handy.

References

Cox, N. J. 2005. Stata tip 17: Filling in the gaps. *Stata Journal* 5: 135–136.

Gould, W. 2003. Stata tip 3: How to be assertive. *Stata Journal* 3: 448.

The Stata Journal (2006)
6, Number 1, pp. 149–150

Stata tip 30: May the source be with you

Nicholas J. Cox
Department of Geography
Durham University
Durham City, UK
n.j.cox@durham.ac.uk

Stata 9 introduced a command, `viewsource`, that does two things: it finds a text file along your ado-path and then opens the Viewer on that text file. See [P] **viewsource** and [U] **17.5 Where does Stata look for ado-files?** for the basic explanations. Naturally, if the text file does not exist as named, say, because you mistyped the name or because it really does not exist, you will get an error message.

Although intended primarily for Stata programmers, `viewsource` can be useful for examining (but not for editing) any text file you are working with. That can include program files, help files, text data files, do-files, log files, and other text documents. Binary or proprietary-format files are not banned, but the command is unlikely to be useful with them. The Viewer is, in particular, not a substitute for any word processor.

Here are some examples. A good way to learn about any Stata command defined by an ado-file is to look at the source code. You might be puzzled by some output, suspect a bug, or simply be curious. Even if you are not (yet) a Stata programmer, you can learn a lot by looking at the code. After all, it is just more Stata. Many, but not all, commands, generically *cmdname*, are defined by *cmdname*.ado—an ado-file with the same name as the command. The exceptions are part of the executable and not visible to you by using `viewsource` or indeed any other command. You might as well start reflexively:

```
. viewsource viewsource.ado
```

from which you will see that `viewsource`'s main actions are to (try to) find the file you specified using `findfile` and then to open the file using `view`. You will see other details too, and puzzling out what they do is a good exercise in program appreciation.

A second example is opening a help file. This action may seem redundant given the existence of the `help` command, but there is a noteworthy exception. `viewsource` fires up `view` with its `asis` option, so that interpreting the SMCL commands in any help file is disabled. This approach is useful for examining the SMCL producing the special markup effects that are evident when you use `help`. Suppose that you see code you want to emulate in your own help files. Then `viewsource` *cmdname*.hlp[1] will show you how that was done, and you do not need to know exactly where the help file is on your machine, except that it must be on your ado-path. A useful template for producing help files is official Stata's `examplehelpfile.hlp`.[1] Again, you do not need to know where it is, or to search for it, as `viewsource` will find it.

---

1. For Stata 10 and later, the extension `.sthlp` is used in place of `.hlp`.

You do not need to be a Stata programmer, or even be interested in the innards of Stata programs or help files, to find `viewsource` useful. Other text files along your ado-path, which includes your current directory or folder, may be opened, too. However, do-files and log files are most likely to be in the current directory or folder, so just using `view` is more direct. Say that you want to repeat a successful but complicated `graph` command, which you carefully stored in a log file or do-file. Use `view` or `viewsource` and then search inside the Viewer using keywords to locate the command before copying it.

Note the emphasis on viewing. The Viewer is not an editor, so making changes to the file is not possible. Use the Viewer when you are clear that neither you nor others should be changing file content, even if the person in question has the requisite file permissions. That situation certainly applies to StataCorp-produced files. With colleagues and students who could do no end of damage if unchecked, this feature is invaluable. It is a limitation if you really do want to edit the file, but then you should already be thinking how to clone `viewsource` so that it fires up your favorite text editor (or Stata's own Do-file Editor). Variants on this idea already exist in Stataland, but writing your own editing command is a good early exercise for any budding Stata programmer.

The Stata Journal (2006)
6, Number 2, pp. 279–280

Stata tip 31: Scalar or variable? The problem of ambiguous names

Gueorgui I. Kolev
Universitat Pompeu Fabra
Barcelona, Spain
gueorgui.kolev@upf.edu

Stata users often put numeric or string values into scalars, which is as easily done as

```
. scalar answer = 42
. scalar question = "What is the answer?"
```

The main discussion of scalars is in [P] **scalar**. Scalars can be used interactively or in programs and are faster and more accurate than local macros for holding values. This advantage would not matter with a number like 42, but it could easily matter with a number that was very small or very large.

Scalars have one major pitfall. It is documented in [P] **scalar**, but users are often bitten by it, so here is another warning. If a variable and a scalar have the same name, Stata always assumes that you mean the variable, not the scalar. This naming can apply even more strongly than you first guess: recall that variable names can be abbreviated so long as the abbreviation is unambiguous (see [U] **11.2 Abbreviation rules**).

Suppose that you are using the `auto` dataset:

```
. sysuse auto
(1978 Automobile Data)
. scalar p = 0.7
. display p
4099
```

What happened to 0.7? Nothing. It is still there:

```
. display scalar(p)
.7
. scalar list p
       p =            .7
```

What is the 4099 result? The dataset has a variable, `price`, and no other variable names begin with `p`, so `p` is understood to mean `price`. Moreover, `display` assumes that if you specify just a variable name, you want to see its value in the first observation, namely, `price[1]`. (The full explanation for that is another story.)

What should users do to be safe?

1. Use a different name. For example, you might introduce a personal convention about uppercase and lowercase. Many Stata users use only lowercase letters (and possibly numeric digits and underscores too) within variable names. Such users

could distinguish scalars by using at least one uppercase letter in their names. You could also use a prefix such as `sc_`, as in `sc_p`. If you forget your convention or if a user of your programs does not know about this convention, Stata cannot sense the interpretation you want. Thus this method is not totally safe.

2. Use the `scalar()` pseudofunction to spell out that you want a scalar. This method is totally safe, but some people find it awkward.

3. Use a temporary name for a scalar, as in

```
. tempname p
. scalar 'p' = 0.7
```

Each scalar with a temporary name will be visible while the program in which it occurs is running and only in that program. Temporary names can be used interactively, too. (Your interactive session also counts for this purpose as a program.) In many ways, this method is the best solution, as it ensures that scalars can be seen only locally, which is usually better programming style.

The Stata Journal (2006)
6, Number 2, p. 281

Stata tip 32: Do not stop

Stephen P. Jenkins
Institute for Social and Economic Research
University of Essex
Colchester, UK
stephenj@essex.ac.uk

The do command, for executing commands from a file, has one (and only one) option: nostop. As the online help file for do indicates, the option "allows the do-file to continue executing even if an error occurs. Normally, Stata stops executing the do-file when it detects an error (nonzero return code)."

This option can be useful in a variety of circumstances. For example,

1. You wish to apply, in a do-file, the same set of commands to data referring to different groups of subjects or different periods. The commands for each dataset might fail with an error because, say, there are no relevant observations or an ml problem may not converge. Running the do-file with the nostop option will allow you to get the desired results for all datasets that do not produce an error and, at the same time, identify the potential source of error in the others. The nostop option is most useful in initial analyses. Error sources in particular datasets, once identified, can be trapped by using capture; if necessary, alternative action may be taken.

2. You have written a command and want to produce a script certifying that using incorrect syntax leads to an appropriate error. There may be several options and several ways in which syntax may be incorrect. With nostop, you can test that each of many incorrectly specified options works as expected, while having the do-file containing the test commands run to completion.

The Stata Journal (2006)
6, Number 2, pp. 282–283

Stata tip 33: Sweet sixteen: Hexadecimal formats and precision problems

Nicholas J. Cox
Department of Geography
Durham University
Durham City, UK
n.j.cox@durham.ac.uk

Computer users generally supply numeric inputs as decimals and expect numerical outputs as decimals. But underneath the mapping from inputs to outputs lies software (such as Stata) and hardware that are really working with binary representations of those decimals. Much ingenuity goes into ensuring that conversions between decimal and binary are invisible to you, but occasionally you may see apparently strange side effects of this fact. This problem is documented in [U] **13.10 Precision and problems therein**, but it still often bites and puzzles Stata users. This tip emphasizes that the special hexadecimal format `%21x` can be useful in understanding what is happening. The format is also documented, but in just one place, [U] **12.5.1 Numeric formats**. Decimal formats such as `%23.18f` can also be helpful for investigating precision problems.

Binary representations of numbers, using just the two digits 0 and 1, can be difficult for people to interpret without extra calculations. The great advantage of a hexadecimal format, using base 16 (i.e., 2^4), is that it is closer to base 10 representations while remaining truthful about what can be held in memory as a representation of a number. It is conventional to use the decimal digits 0–9 and the extra digits a–f when base 16 is used. Thus a represents 10 and f represents 15. Hence, at its simplest, hexadecimal 10 represents decimal 16, hexadecimal 11 represents decimal 17, and so forth. (Think of 11 as $1 \times 16^1 + 1 \times 16^0$, for example.) In practice, we want to hold fractions and, as far as possible, some extremely large and extremely small numbers. The general format of a hexadecimally represented number in Stata is thus $m\mathrm{X}p$, to be read as $m \times 2^p$. Thus if you use the format `%21x` with `display`, you can see examples:

```
. di %21x 1
+1.0000000000000X+000
. di %21x -16
-1.0000000000000X+004
. di %21x 1/16
+1.0000000000000X-004
```

You see that 1, -16, and $1/16$ are, respectively, 1×2^0, -1×2^4, and 1×2^{-4}.

The special format is useful to others besides the numerical analysts mentioned in [U] **12.5.1 Numeric formats**. If you encounter puzzling results, looking at the numbers in question should help clarify what Stata is doing and why it does not match your expectation.

Users get bitten in two main ways. First, they forget that most of the decimal digits .1, .2, .3, .4, .5, .6, .7, .8, and .9 cannot be held exactly. Of these, only .5 (1/2) can possibly be represented exactly by a binary approximation; all the others must be held approximately only—regardless of how many bytes are used. To convince yourself of this, see that, e.g., 42.5 can be held exactly,

```
. di %21x 42.5
+1.5400000000000X+005
. di (1 + 5/16 + 4/256) * 2^5
42.5
```

whereas 42.1 cannot be held exactly,

```
. di %21x 42.1
+1.50ccccccccccdX+005
. di %23.18f 42.1
  42.100000000000001421
```

Close, but not exact. Second, users forget that although very large or very small numbers can be held approximately, not all possible numbers can be distinguished, even when those numbers are integers within the limits of the variable type being used.

A common source of misery is trying to hold nine-digit integers in numeric variables. If these are identifiers, holding them as str9 variables is a good idea, but let us focus on what often happens when users read such integers into numeric variables. This experiment shows the problems that can ensue.

```
. gen pinid = 123456789
. di %9.0f pinid[1]
123456792
. di %21x pinid[1]
+1.d6f3460000000X+01a
```

Stata did not complain, but it did not oblige. The value is off by 3. You will see that the value held is a multiple of 4, as the last two digits 92 are divisible by 4. Did we or Stata do something stupid? Can we fix it?

```
. replace pinid = pinid - 3
(0 real changes made)
```

Trying to subtract 3 gives us the same number, so far as Stata is concerned. What is going on? By default, Stata is using a float variable. See [D] **data types** if you want more information. At this size of number, such a variable can hold only multiples of 4 exactly, so we lose many final digits. The remedy, if a numeric variable is needed, is to use a long or double storage type instead.

The Stata Journal (2006)
6, Number 3, pp. 425–427

Stata tip 34: Tabulation by listing

David A. Harrison
Intensive Care National Audit and Research Centre
London, UK
david.harrison@icnarc.org

The command `list` is often regarded as simply a data management tool for listing observations, but it has several little-used options that make it a useful tool for producing customized tables.

The dataset `auto.dta` contains the 1978 repair records (rated 1–5) for various makes of car. We can use `list` in its typical manner to look at some of the data:

```
. use make rep78 using http://www.stata-press.com/data/r9/auto
(1978 Automobile Data)

. drop if missing(rep78)
(5 observations deleted)

. list in 1/5
```

	make	rep78
1.	AMC Concord	3
2.	AMC Pacer	3
3.	Buick Century	3
4.	Buick Electra	4
5.	Buick LeSabre	3

Suppose that we wish to tabulate the makes of car according to repair record. There is no simple approach to produce tables containing strings; however, if we can modify the data so that the variables in the new dataset represent the columns of our desired table and the observations represent the rows, then we can produce the table with a `list` command. The `reshape` command (see [D] **reshape**) provides the ideal tool to do this:

```
. by rep78 (make), sort: gen row = _n

. reshape wide make, i(row) j(rep78)
(note: j = 1 2 3 4 5)
```

Data	long	->	wide
Number of obs.	69	->	30
Number of variables	3	->	6
j variable (5 values)	rep78	->	(dropped)
xij variables:			
	make	->	make1 make2 ... make5

The first command above generated a variable defining the rows of our table. Sorting on `make` will ensure that the makes of car appear in alphabetical order within the table. After this `reshape`, the results of a simple `list` will still not be ideal, as the columns will be headed `make1`, `make2`, etc. We can change this by using the `subvarname` option, which substitutes the characteristic `varname` for each variable as the column heading. Characteristics (see [P] **char**) are named items of text that can be attached to any variable or to the entire dataset. We use `forvalues` to loop through the values of `rep78` creating these characteristics. A previous article in the *Stata Journal* has discussed more complex looping, including using the `levelsof` (previously `levels`) and `foreach` commands to cycle through all values of a variable (Cox 2003). Other options for `list` remove observation numbers, remove the default horizontal lines every five rows, insert dividers between the columns, and make the columns of equal width:

```
. forvalues i = 1/5 {
  2.     char make`i'[varname] "Repair record `i'"
  3. }
. list make1-make3, noobs sep(0) divider nocompress subvarname
```

Repair record 1	Repair record 2	Repair record 3
Olds Starfire	Cad. Eldorado	AMC Concord
Pont. Firebird	Chev. Monte Carlo	AMC Pacer
	Chev. Monza	Audi Fox
	Dodge Diplomat	Buick Century
	Dodge Magnum	Buick LeSabre
	Dodge St. Regis	Buick Regal
	Plym. Volare	Buick Riviera
	Pont. Sunbird	Buick Skylark
		Cad. Deville
		Cad. Seville
		Chev. Chevette
		Chev. Malibu
		Chev. Nova
		Fiat Strada

(*output omitted*)

Only the first three columns are displayed here because of the width of the page, but if you require more columns than can be displayed in your results window (and are logging your output), you can use the `linesize(#)` option to increase the available width.

Two-way tables can be achieved in a similar manner:

```
. use make rep78 foreign using http://www.stata-press.com/data/r9/auto
(1978 Automobile Data)
. drop if missing(rep78)
(5 observations deleted)
. by rep78 foreign (make), sort: gen row = _n
. qui reshape wide make, i(rep78 row) j(foreign)
. gen str1 rep78txt = string(rep78) if row == 1
(50 missing values generated)
. format rep78txt %-1s
. char rep78txt[varname] "Repair record"
. char make0[varname] "Domestic"
. char make1[varname] "Foreign"
```

```
. list rep78txt make0 make1, noobs sepby(rep78) div noc subvar abbrev(13)
```

Repair record	Domestic	Foreign
1	Olds Starfire Pont. Firebird	
2	Cad. Eldorado Chev. Monte Carlo Chev. Monza Dodge Diplomat Dodge Magnum Dodge St. Regis Plym. Volare Pont. Sunbird	
3	AMC Concord AMC Pacer Buick Century Buick LeSabre Buick Regal	Audi Fox Fiat Strada Renault Le Car

(*output omitted*)

Do not underestimate what can be achieved with a simple `list`!

Reference

Cox, N. J. 2003. Speaking Stata: Problems with lists. *Stata Journal* 3: 185–202.

The Stata Journal (2006)
6, Number 3, pp. 428–429

Stata tip 35: Detecting whether data have changed

William Gould
StataCorp
College Station, TX
wgould@stata.com

Included in the 17 May 2006 update is a new command that you may find useful, `datasignature`. If you have installed the update, type `help datasignature`. If you have not updated, or are unsure, type `update query` to find out, and type `update all` to install.

Here is the result of running `datasignature` on `auto.dta`:

```
. sysuse auto
(1978 automobile data)
. datasig
   74:12(71728):3831085005:1395876116
```

The output is `auto.dta`'s data signature. If you change the data, even just a little bit, the last two numbers will change:

```
. replace mpg = mpg + 1 in 2
(1 real change made)
. datasig
   74:12(71728):1616229321:1400086868
```

If you change the name of a variable, the last two numbers stay the same and the number inside the parentheses changes:

```
. rename mpg miles_per_gallon
. datasig
   74:12(57876):1616229321:1400086868
```

`datasignature` is designed to help those who use data maintained by others and those who worry that they might themselves have accidentally changed their data.

In the latter case, you could save the signature in the dataset,

```
. datasig
   74:12(57876):1616229321:1400086868
. note: 'r(datasignature)'
```

and check it later,

```
. notes
_dta:
  1.  from Consumer Reports with permission
  2.  74:12(57876):1616229321:1400086868
```

Another idea is to include `datasignature` at the beginning of logs:

```
─────────────────────────── begin myfile.do ───
log using myfile, replace
sysuse auto, clear
datasig

...
log close
─────────────────────────── end myfile.do ───
```

You can specify a varlist and `if` and `in`, so if you have a large dataset and want to check just part of it, you can write

```
─────────────────────────── begin myfile.do ───
log using myfile, replace
sysuse auto, clear
datasig mpg weight price if foreign

...
log close
─────────────────────────── end myfile.do ───
```

The Stata Journal (2006)
6, Number 3, pp. 430–432

Stata tip 36: Which observations?

Nicholas J. Cox
Department of Geography
Durham University
Durham City, UK
n.j.cox@durham.ac.uk

A common question is how to identify which observations satisfy some specified condition. The easiest answer is often to use `list`, as in

```
. use http://www.stata-press.com/data/r9/auto, clear
(1978 Automobile Data)
. list rep78 if rep78 == 3
  (output omitted)
```

An equivalent is to use `edit` instead. In either case, the basic ingredients to an answer are

1. At least an `if` condition and possibly an `in` condition, too. Even if we start out interested in all observations, the condition of interest will be specified using `if`.

2. The observation numbers themselves. Evidently some commands will show them (`list` and `edit` being examples), but otherwise we will need to work a little harder and do something like

   ```
   . gen long obsno = _n
   ```

   and work with that new variable. Here I spelled out that the variable type to be used is a `long`. Consulting the help for data types shows that an `int` will work for datasets with up to 32,740 observations. The default for a new variable is `float`: this will often be fine, but it is dangerous for very large datasets because not every large integer less than Stata's maximum dataset size can be held exactly.

What other complications will we need to worry about when specifying conditions?

- Precision problems with noninteger values, prominently documented but nevertheless a frequent source of minor grief (e.g., see Cox [2006] and references therein).

- Ties; i.e., more than one observation may satisfy a specified condition.

- Conditions involving string comparisons as well as numeric comparisons.

`list` or `edit` shows us the observation numbers for a particular condition, but not compactly or retrievably. We do not want to have to type out those numbers if we need them for some other purpose. To get a more compact display, one approach uses `levelsof` after generating an observation number variable.

```
. levelsof obsno if rep78 == 3
1 2 4 6 8 9 10 11 13 14 16 19 25 26 27 28 31 32 34 36 37 39 41 42 44 49 50 54 60 65
```

In an (updated) Stata 8, use `levels`, not `levelsof`. The help for `levelsof` shows that you can put the list of observation numbers into a local macro for further manipulation and that this list is accessible immediately after issuing the command as `r(levels)`.

If you want the `obsno` variable for this kind of purpose, you might want it shortly for something similar, so it might as well be left in memory as long as there is plenty to spare. But `obsno` will remain identical in contents to `_n` only as long as the sort order is not changed.

```
. assert obsno == _n
```

is a good way to check whether that remains true. `assert` gives no output if the assertion made is true for every observation, no news thus being good news in this example. See also Gould (2003).

Asking for the levels of an observation number variable works when ties are present and when string comparisons are specified. You can also add whatever other `if` or `in` conditions apply.

The main problem to worry about in practice is the precision problem. Consider

```
. summarize gear
```

Variable	Obs	Mean	Std. Dev.	Min	Max
gear_ratio	74	3.014865	.4562871	2.19	3.89

What if we want to see which observations are equal to the maximum?

```
. levelsof obsno if gear == 3.89
```

shows nothing and so fails to find the observation(s), whereas

```
. levelsof obsno if gear == float(3.89)
56
```

happens to give the right answer, but you will not always be so lucky. In other circumstances, what you see (3.89) might be more rounded than it should be. The best approach in general is to use the saved results produced by commands such as those, which are documented in the manual entry for each command. Thus after `summarize`,

```
. levelsof obsno if gear == r(max)
56
```

gives the right answer, as it does in this example,

```
. levelsof obsno if gear == `r(max)'
56
```

Nevertheless, I recommend using `r(max)` rather than '`r(max)`' because the former gives you access to the maximum precision possible. A similar comment applies to e-class results.

Incidentally, because `levelsof` is r-class it will overwrite the r-class results left behind by `summarize`, so you will need to issue such commands in the right order. Thus if we wanted to see both the maximums and the minimums, we would need to repeat commands. As a variation, we use the `meanonly` option, which despite its name does leave the maximum and minimum in memory.

```
. summarize gear, meanonly
. levelsof obsno if x == r(max)
56
. summarize x, meanonly
. levelsof obsno if x == r(min)
12
```

References

Cox, N. J. 2006. Stata tip 33: Sweet sixteen: Hexadecimal formats and precision problems. *Stata Journal* 6: 282–283.

Gould, W. 2003. Stata tip 3: How to be assertive. *Stata Journal* 3: 448.

The Stata Journal (2006)
6, Number 4, pp. 588–589

Stata tip 37: And the last shall be first

Christopher F. Baum
Department of Economics
Boston College
Chestnut Hill, MA 02467
baum@bc.edu

Mata's built-in function list contains many useful matrix operations, but I recently came upon one that was lacking: the ability to *flip* a matrix along its rows or columns. Either of those operations can readily be done as a Mata statement, but I'd rather not remember the syntax—or have to remember what it is meant to do when I reread the code. So I wrote these two simple functions:[1]

```
mata:
matrix function flipud(matrix X) {
        return(rows(X)>1 ? X[rows(X)..1,.] : X)
}


matrix function fliplr(matrix X) {
        return(cols(X)>1 ? X[.,cols(X)..1] : X)
}
end
```

These functions will flip a matrix `ud`—upside down (the first row becomes the last)— or `lr`, left to right (the first column becomes the last). Because the functions take a `matrix` argument, they may be applied to any of Mata's matrix types, including `string` matrices.

Users have asked why one would want to flip a matrix "upside down". As it happens, doing so becomes a handy tool when creating a two-sided linear filter. Say that we have defined a vector x, containing a declining set of weights: a one-sided linear filter. We can turn x into a two-sided set of weights by using `flipud()`:

```
. mata:
                                              ------ mata (type end to exit) ------
: x = (1\0.5\0.25\0.125\0.0625) ; x
            1

    1       1
    2       .5
    3       .25
    4       .125
    5       .0625
```

---

1. I thank Mata's principal architect, William Gould, for improvements he suggested that make the
code more general.

```
: x = (flipud(x[2..rows(x)]) \ x); x
             1

  1  |   .0625
  2  |   .125
  3  |   .25
  4  |   .5
  5  |   1
  6  |   .5
  7  |   .25
  8  |   .125
  9  |   .0625

: end
```

To decipher that statement, note that `2..rows(x)` refers to the second through last rows of vector `x`. The statement thus flips those rows of `x` upside down and concatenates them to the original `x` by using the *column-join* operator (see [M-2] **op_join**).

As a second example, consider applying both functions to a string matrix:

```
. mata:
──────────────────────────────────────────────── mata (type end to exit) ───────
: Greek2me = ("alpha","beta","gamma"\"delta","epsilon","zeta"\"eta","theta",
> "iota"\"kappa","lambda","mu"\"nu","xi","omicron"\"pi",
> "rho","sigma"\"tau","upsilon","phi"\"chi","psi","omega")
: Greek2me
               1           2           3

  1  |     alpha        beta       gamma
  2  |     delta     epsilon        zeta
  3  |       eta       theta        iota
  4  |     kappa      lambda          mu
  5  |        nu          xi     omicron
  6  |        pi         rho       sigma
  7  |       tau     upsilon         phi
  8  |       chi         psi       omega

: lastFirst = fliplr(flipud(Greek2me)); lastFirst
               1           2           3

  1  |     omega         psi         chi
  2  |       phi     upsilon         tau
  3  |     sigma         rho          pi
  4  |   omicron          xi          nu
  5  |        mu      lambda       kappa
  6  |      iota       theta         eta
  7  |      zeta     epsilon       delta
  8  |     gamma        beta       alpha

: end
```

The Stata Journal (2006)
6, Number 4, pp. 590–592

Stata tip 38: Testing for groupwise heteroskedasticity

Christopher F. Baum
Department of Economics
Boston College
Chestnut Hill, MA 02467
baum@bc.edu

A natural source of heteroskedasticity in many kinds of data is *group membership*: observations in the sample may be a priori defined as members of groups, and the variance of a series may differ considerably across groups. This concept will also apply to the errors from a linear regression. The assumption of homoskedasticity in the relationship may reasonably hold within each group, but not between groups. This assumption most commonly arises in cross-sectional datasets. In economic data, for instance, the groups may correspond to firms in different industries or workers in different occupations. It could also apply in a time-series context: for instance, the variance of daily temperature may not be constant over the four seasons. In any case, a test for heteroskedasticity of this sort should take this a priori knowledge into account.

How might we test for groupwise heteroskedasticity in a variable or in the errors from a regression? In the context of regression, if we can argue that each group's regression equation satisfies the classical assumptions (including that of homoskedasticity), the s^2 computed by regress (see [R] **regress**) is a consistent estimate of the group-specific variance of the disturbance process. For two groups, an F test may be constructed, with the larger variance in the numerator; the degrees of freedom are the residual degrees of freedom of each group's regression. Conducting an F test is easy if both groups' residuals are stored in one variable, with a group variable indicating group membership (in this case 1 or 2). The third form of sdtest may then be used, with the by(*groupvar*) option, to conduct the F test.

What if there are more than two groups across which we wish to test for equality of disturbance variance, for instance, a set of 10 industries? We may then use the robvar command (see [R] **sdtest**), which like sdtest expects to find one variable containing each group's residuals, with a group membership variable identifying them. The by(*groupvar*) option is used here as well. The test conducted is that of Levene (1960) labeled as W_0, which is robust to nonnormality of the error distribution. Two variants of the test proposed by Brown and Forsythe (1974), which uses more robust estimators of central tendency (e.g., median rather than mean), W_{50} and W_{10}, are also computed.

We illustrate groupwise heteroskedasticity with state-level data: 1 observation per year for each of the six states in the New England region of the United States for 1981–2000. We first apply robvar to the state-level population series to examine whether the variance of population is constant across states.

```
. use http://www.stata-press.com/data/imeus/NEdata
. robvar pop, by(state)
```

	Summary of pop		
state	Mean	Std. Dev.	Freq.
CT	3276614.5	81452.212	20
MA	6030915.5	178354.76	20
ME	1212718.1	46958.538	20
NH	1094238.9	94362.302	20
RI	1000209.9	29548.701	20
VT	562960.65	31310.625	20
Total	2196276.3	1931629.4	120

```
W0  =  13.856324   df(5, 114)    Pr > F = 0.00000000
W50 =  11.820938   df(5, 114)    Pr > F = 0.00000000
W10 =  13.306895   df(5, 114)    Pr > F = 0.00000000
```

All forms of the test clearly reject the hypothesis of homoskedasticity across states' population series: hardly surprising when the standard deviation of Massachusetts' (MA) population is six times that of Rhode Island (RI).

We now fit a linear trend model to state disposable personal income per capita, dpipc, by regressing that variable on year. The residuals are tested for equality of variances across states with robvar.

```
. regress dpipc  year
```

Source	SS	df	MS
Model	3009.33617	1	3009.33617
Residual	806.737449	118	6.83675804
Total	3816.07362	119	32.0678456

```
Number of obs =     120
F( 1,   118) =  440.17
Prob > F      =  0.0000
R-squared     =  0.7886
Adj R-squared =  0.7868
Root MSE      =  2.6147
```

dpipc	Coef.	Std. Err.	t	P>|t|	[95% Conf. Interval]
year	.8684582	.0413941	20.98	0.000	.7864865 .9504298
_cons	−1710.508	82.39534	−20.76	0.000	−1873.673 −1547.343

```
. predict double eps, residual
. robvar eps, by(state)
```

	Summary of Residuals		
state	Mean	Std. Dev.	Freq.
CT	4.167853	1.3596266	20
MA	1.618796	.86550138	20
ME	−2.9841056	.93797625	20
NH	.51033312	.61139299	20
RI	−.8927223	.63408722	20
VT	−2.4201543	.71470977	20
Total	−6.063e−14	2.6037101	120

```
W0  =  4.3882072   df(5, 114)    Pr > F = 0.00108562
W50 =  3.2989851   df(5, 114)    Pr > F = 0.00806751
W10 =  4.2536245   df(5, 114)    Pr > F = 0.00139064
```

The hypothesis of equality of variances is soundly rejected by all three `robvar` test statistics, with the residuals for Connecticut, Massachusetts, and Maine possessing a standard deviation considerably larger than those of the other three states.

References

Brown, M. B., and A. B. Forsythe. 1974. Robust tests for the equality of variances. *Journal of the American Statistical Association* 69: 364–367.

Levene, H. 1960. Robust tests for equality of variances. In *Contributions to Probability and Statistics: Essays in Honor of Harold Hotelling*, ed. I. Olkin, S. G. Ghurye, W. Hoeffding, W. G. Madow, and H. B. Mann, 278–292. Menlo Park, CA: Stanford University Press.

The Stata Journal (2006)
6, Number 4, pp. 593–595

Stata tip 39: In a list or out? In a range or out?

Nicholas J. Cox
Department of Geography
Durham University
Durham City, UK
n.j.cox@durham.ac.uk

Two simple but useful functions, inlist() and inrange(), were added in Stata 7, but users somehow still often overlook them. The manual entry [D] **functions** gives formal statements on definitions and limits. The aim here is to emphasize with examples how natural and helpful these functions can be.

The question answered by inlist() is whether a specified argument belongs to a specified list. That answered by inrange() is whether a specified argument falls in a specified range. We can ask the converse question, of not belonging to or falling outside a list or range, by simply negating the function. Thus !inlist() and !inrange() can be read as "not in list" and "not in range".

These functions can reduce your typing, reduce the risk of small errors, and make your Stata code easier to read and maintain. Thus with the auto data in memory, consider the choice for the integer-valued variable rep78 between older ways of getting a simple listing,

```
. list make rep78 if rep78 == 3 | rep78 == 4 | rep78 == 5
. list make rep78 if rep78 >= 3 & rep78 <= 5
. list make rep78 if rep78 > 2 & rep78 < 6
```

and newer ways of getting the same listing,

```
. list make rep78 if inlist(rep78, 3, 4, 5)
. list make rep78 if inrange(rep78, 3, 5)
```

The examples here are typical of a good way to use inlist() or inrange(): move directly from feeding arguments to each function to using the results of the calculation. If you wanted to keep the results, you could put them into a variable (or a macro). The result of inlist() or inrange() is either 1 when the value specified is in range or in list and 0 otherwise (and thus never missing). So, if you use a variable to store results, let it be a byte variable for efficiency in storage.

In more detail: so long as none of the arguments z, a, b is missing, $inrange(z, a, b)$ is true whenever $z \geq a$ and $z \leq b$. Thus inrange(60, 50, 70) is true (numerically 1) because $60 \geq 50$ and $60 \leq 70$. However, inrange(60, 70, 50) is false (0) because 60 is not ≥ 70 and 60 is not ≤ 50. Thus the order of a and b is crucial. There are situations when you are not sure in advance about the ordering of arguments, but you can always use devices such as inrange(z, min(a, b), max(a, b)) (which tests whether one value is between two others).

The definition of inrange() is more complicated when any argument is numeric missing. See [D] **functions** for the precise definitions. The most important example is that `inrange(`z`, ` a`, ` `.)` is interpreted as $z \geq a$ and $z < .$ (z greater than or equal to a, but not missing). This may look like a bug, but it is really a feature. Even experienced users sometimes forget that in Stata numeric missing is regarded as arbitrarily large. Hence, `z >= 42` will be true for all the missing values of `z`, as well as for all values that are greater than or equal to 42. The longstanding workaround when this is not what you want with regard to missing values is to add the extra condition that `z` is not missing, as in `z >= 42 & z < .`, but `inrange(z, 42, .)` is another way to do this.

The definitions that come into play when any argument is missing imply that `inrange()` is not a good tool to use when you want to test for numeric missings (including any comparisons with extended missing values). For that it is better to use `missing()`, `inlist()`, or combined statements using simple inequalities.

`inlist()` and `inrange()` can often be used with the in-built quantities _n and _N specifying, respectively, the current observation number and the current number of observations. Sometimes users wish to specify that a command should apply to an irregular set of observation numbers, and `if inlist(_n,17,42,99,217)` exemplifies how that could be done with a small set (the limit is 255 numbers and is unlikely to bite in sensible practice). A pitfall here is clearly that any sorting of the dataset will often imply that the observations concerned end up in different positions. Thus saving the results of this computation in a byte variable will often be a good idea. This approach is not better general practice than using criteria such as those based on variable values, but there may be occasions when you will want this feature.

Other examples of the same kind arise with longitudinal or panel data. Recently I wanted to identify the first and last values of a response in each panel, and

```
. by panelvar (timevar): gen y_ends = y if inlist(_n, 1, _N)
```

offers a way to do that. Conversely, `!inlist(_n, 1, _N)` identifies all the others. Whether you prefer that `if` condition to the more traditional `if _n == 1 | _n == _N` is admittedly a matter of taste. Using `in` is not an option here because `in` may not be combined with `by:`.

The examples so far are all for numeric arguments. The arguments of either function can be all numeric or all string. Thus given one character, c, `inrange("`c`", "a", "z")` tests whether c is one of the 26 lowercase letters of the English alphabet; correspondingly, `inrange("`c`", "A", "Z")` tests whether c is one of the 26 uppercase letters of the same alphabet. More generally, `inrange("`*string*`", "a", "z")` tests whether *string* begins with a lowercase letter, and correspondingly for the arguments `"A"`, `"Z"` and uppercase letters. Because lowercase and uppercase letters are typically not adjacent in your computer's character sets, be careful when working with both. If you were indifferent about the distinction between uppercase and lowercase, you could work with `lower("`*string*`)"` or `upper("`*string*`")`.

These examples lead directly to a way of filtering a string variable to select characters that you want or ignore characters that you don't. Suppose that we wanted to select only the alphabetic characters in a string variable. Check the variable type to see its maximum length (18, or whatever), `generate` a new empty-string variable, and then loop over the characters, adding them to the end of the new variable only if they are as desired.

```
generate newvar = ""
quietly forvalues i = 1/18 {
        replace newvar = newvar + substr(oldvar,`i',1) ///
                if inrange(lower(substr(oldvar,`i',1)),"a","z")
}
```

Commands like this tend to become rather long, but they are not in principle complicated. The attraction of a low-level approach is that you can design exactly the filter you wish according to the precise problem it is intended to solve.

A further simple but general moral evident from various examples here is that the power of Stata functions often arises from how they can be combined.

The Stata Journal (2006)
6, Number 4, p. 596

Stata tip 36: Which observations? Erratum

Nicholas J. Cox
Department of Geography
Durham University
Durham City, UK
n.j.cox@durham.ac.uk

In the last code example,

```
. summarize gear, meanonly
. levelsof obsno if x == r(max)
56
. summarize x, meanonly
. levelsof obsno if x == r(min)
12
```

the last three commands should refer to the variable gear, not to x.

The Stata Journal (2007)
7, Number 1, pp. 137–139

Stata tip 40: Taking care of business

Christopher F. Baum
Department of Economics
Boston College
Chestnut Hill, MA 02467
baum@bc.edu

Daily data are often generated by nondaily processes: for instance, financial markets are closed on weekends and holidays. Stata's time-series date schemes ([U] **24.3 Time-series dates**) allow for daily data, but gaps in time series may be problematic. A model that uses lags or differences will lose several observations every time a gap appears, discarding many of the original data points. Analysis of "business-daily" data often proceeds by assuming that Monday follows Friday, and so on. At the same time, we usually want data to be placed on Stata's time-series calendar so that useful tools such as the `tsline` graph will work and label data points with readable dates; see [TS] **tsline**.

At a recent Stata Users Group presentation in Boston, David Drukker spoke to this point. His solution: generate two date variables, one containing the actual calendar dates, another numbering successive available observations consecutively. The former variable (`caldate`) is `tsset` (see [TS] **tsset**) when the calendar dates are to be used, whereas the latter (`seqdate`) is `tsset` when statistical analyses are to be performed.

We download daily data on the 3-month U.S. Treasury bill rate with Drukker's `freduse` command (Drukker 2006) and retain the August 2005–present data for analysis. (We can also view the data graphically with `tsline`.)

```
. freduse DTB3
(13764 observations read)
. rename daten caldate
. tsset caldate
        time variable:  caldate, 04jan1954 to 05oct2006, but with gaps
. keep if tin(1aug2005,)
(13455 observations deleted)
. label var caldate date
```

These data do not contain observations for weekends and are missing for U.S. holidays. We may not want to drop the observations containing missing data, though, as we may have complete data for other variables: for instance, exchange rate data are available every day. If there were no missing data in our series—only missing observations—we could use Drukker's suggestion and `generate seqdate = _n`. As we have observations for which `DTB3` is missing, we follow a more complex route:

```
. quietly generate byte notmiss = DTB3 < .
. quietly generate seqdate = cond(notmiss, sum(notmiss),.)
. tsset seqdate
        time variable:  seqdate, 1 to 297
```

The variable `seqdate` is created as the sequential day number for every nonmissing day and is itself missing when DTB3 is missing—allowing us to use this variable in `tsset` and then use time-series operators (see [U] **11.1.1 varlist**) in `generate` or estimation commands such as `regress`. We may want to display the transformed data (or results from estimation, such as predicted values) on a time-series graph. We can just revert to the other `tsset`:

```
. quietly generate dDTB3 = D.DTB3
. quietly regress dDTB3 L(1/5).dDTB3
. predict double dDTB3hat, xb
(18 missing values generated)
. label var dDTB3 "Daily change in three-month Treasury rate"
. tsset caldate
        time variable:  caldate, 01aug2005 to 05oct2006, but with gaps
. tsline dDTB3 dDTB3hat, yline(0) xlabel(, angle(45))
```

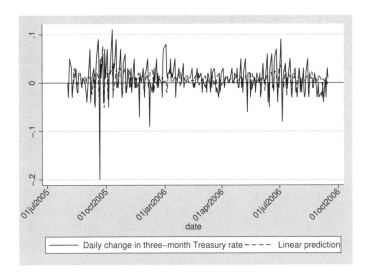

If we retain both the `caldate` and `seqdate` variables in our saved dataset, we will always be able to view these data either on a time-series calendar or as a sequential series. In my research, I need to know how many calendar days separate each observed point (1 for Thursday–Friday but 3 for Friday–Monday) and then sum DTB3 by month, weighting each observation by the square root of the days of separation:

```
. tsset seqdate
        time variable:  seqdate, 1 to 297
. quietly generate dcal = D.caldate if seqdate < .
. quietly generate month = mofd(caldate) if seqdate < .
. format %tm month
. sort month (seqdate)
. quietly by month: generate adjchange = sum(dDTB3/sqrt(dcal))
. quietly by month: generate sumchange = adjchange if _n==_N & month < .
```

```
. list month sumchange if sumchange < ., sep(0) noobs
```

month	sumchange
2005m8	-.003812
2005m9	-.0810769
2005m10	.2424316
2005m11	-.063453
2005m12	.096188
2006m1	.2769615
2006m2	.099641
2006m3	.0142265
2006m4	.0938675
2006m5	.0350555
2006m6	.0327906
2006m7	.0304485
2006m8	-.083812
2006m9	-.123094
2006m10	.0442265

Reference

Drukker, D. M. 2006. Importing Federal Reserve economic data. *Stata Journal* 6: 384–386.

The Stata Journal (2007)
7, Number 1, p. 140

Stata tip 41: Monitoring loop iterations

David A. Harrison
Intensive Care National Audit & Research Centre
London, UK
david.harrison@icnarc.org

If, like me, you have ever started a Stata program running and returned hours later to find it still running with no idea of whether it is actually getting anywhere, then you will be looking for a simple method to monitor your loops. The commands `bootstrap` and `jackknife` produce an attractive table of dots for just this purpose by using the undocumented command `_dots`.

Any loop can easily be modified to report its progress by using `_dots`:

```
nois _dots 0, title(Loop running) reps(100)
forvalues i = 1/100 {
    (main body of loop)
    nois _dots `i' 0
}

Loop running (100)
    ———+— 1 ——+— 2 ——+— 3 ——+— 4 ——+— 5
.................................................    50
.................................................   100
```

The first `_dots` command, called with the argument 0, sets up the graduated header line. The title and number of repetitions are optional. Further calls to `_dots` take two arguments: the repetition number and a return code. The return code 0 (as used above) indicates a successful repetition, and a dot is displayed. Alternative return codes produce a green 's' (-1) or a red 'x' (1), 'e' (2), 'n' (3) or '?' (any other value).

Below is a more complicated example using a `while` loop. Here, the loop runs until 70 successes are achieved. For this contrived example, each iteration succeeds at random with 80% probability. Successes are reported with a dot (`.`) and failures with an x.

```
nois _dots 0, title(Looping until 70 successes...)
local rep 1
local nsuccess 0
while `nsuccess' < 70 {
    local fail = uniform() < .2
    local nsuccess = `nsuccess' + (`fail' == 0)
    nois _dots `rep++' `fail'
}

Looping until 70 successes...
    ———+— 1 ——+— 2 ——+— 3 ——+— 4 ——+— 5
xx...x..x...xx...xx....x...............x.........    50
.x.............x..x.x.x....x.......
```

The Stata Journal (2007)
7, Number 1, pp. 141–142

Stata tip 42: The overlay problem: Offset for clarity

James Cui
Department of Epidemiology and Preventive Medicine
Monash University
Melbourne, Australia
james.cui@med.monash.edu.au

A common graphical problem often arises when one graph axis shows a discrete scale and the other shows a continuous scale. The discrete scale could, for example, represent distinct categories or a series of times at which data were observed. If we want to show several quantities on the continuous axis, matters may easily become confused—and confusing—when some of those quantities are close, especially if they are shown as confidence or other intervals. One answer to this overlap problem is to offset for clarity.

For example, in longitudinal studies, we often need to draw the mean response and 95% confidence intervals of a continuous variable for several categories over the follow-up period. However, the confidence intervals can overlap if the difference between the mean responses is small. Consider an example closely based on one in Rabe-Hesketh and Everitt (2007, 144–166). Mean and standard deviation of depression score, `dep` and `sddep`, have been calculated for each of five visits and two treatment groups, `visit` and `group`. The number of subjects in each combination of visit and group is also given as `n`, so that approximate 95% confidence limits `high` and `low` can be based on twice the standard error, $\text{sddep}/\sqrt{n}$. See table 1.

Table 1: Mean and standard deviation of depression score over visit

visit	group	dep	sddep	n	high	low
1	Placebo	16.48	5.28	27	18.51	14.45
1	Estrogen	13.37	5.56	34	15.28	11.46
2	Placebo	15.89	6.12	22	18.50	13.28
2	Estrogen	11.74	6.58	31	14.10	9.38
3	Placebo	14.13	4.97	17	16.54	11.72
3	Estrogen	9.13	5.48	29	11.17	7.09
4	Placebo	12.27	5.85	17	15.11	9.43
4	Estrogen	8.83	4.67	28	10.60	7.06
5	Placebo	11.40	4.44	17	13.55	9.25
5	Estrogen	7.31	5.74	28	9.48	5.14

To plot these results, we first use `clonevar` to make a copy of `visit` as `x`: that way, `x` inherits format and value labels as well as values from `visit`, not important here but useful in other problems. We copy so that the original `visit` remains unchanged. Adding and subtracting a small value depending on `group` offsets the two groups. Clearly, the value here, 0.05, can be varied according to taste. If there had been three groups, we could have left one where it was and moved the other two. Because the number of groups is either even or odd, a symmetric placement around integer values on the discrete axis can thus be achieved either way.

```
. use depression

. clonevar x = visit

. replace x = cond(group == "Placebo", x - 0.05, x + 0.05)
x was byte now float
(10 real changes made)

. twoway (connected dep x if group == "Placebo", lpattern(solid) msymbol(D))
>        (connected dep x if group == "Estrogen", lpattern(dash) msymbol(S))
>        (rcap high low x if group == "Placebo")
>        (rcap high low x if group == "Estrogen")
>        , xlabel(1 2 3 4 5) ylab(5(5)20, format(%5.0f))
>        xtitle("Visit") ytitle("Depression score")
>        legend(pos(1) ring(0) col(1) order(1 "Placebo" 2 "Estrogen"))
```

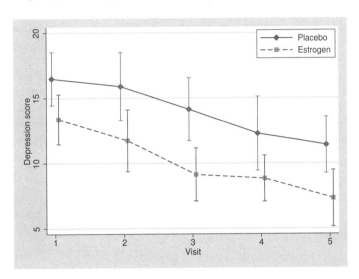

Figure 1: Mean depression score and 95% confidence intervals over visit

Reference

Rabe-Hesketh, S., and B. Everitt. 2007. *A Handbook of Statistical Analyses Using Stata*. 4th ed. Boca Raton, FL: Chapman & Hall/CRC.

The Stata Journal (2007)
7, Number 1, pp. 143–145

Stata tip 43: Remainders, selections, sequences, extractions: Uses of the modulus

Nicholas J. Cox
Department of Geography
Durham University
Durham City, UK
n.j.cox@durham.ac.uk

The `mod(x, y)` function produces remainders from division. It yields the remainder or residue when `x` is divided by `y`. The manual or online help definition is that `mod(x, y)` yields the modulus of `x` with respect to `y`. Mathematically, this definition is an abuse of terminology, but one that Stata shares with many other computing languages. In mathematics, the modulus is the divisor; somehow a few decades back in computing the term was transferred to the remainder.

Like several other functions, `mod()` may at first seem fairly trivial, so here are examples of some of its uses. All illustrations will be for first arguments (dividends) that are zero or positive integers and second arguments (divisors) that are positive integers. Stata's definition is more general, and yet more general definitions are possible, but the illustrations will show the main idea and cover most practical applications. Texts on discrete mathematics or the mathematics behind computing give fuller treatments (Biggs 2002; Knuth 1997; Graham, Knuth, and Patashnik 1994), but we need none of that material here. Authors often discuss these ideas under the heading of congruences.

How should you play with functions like `mod()` to get to know them? First, there is `display`:

```
. display mod(1,2)
1
. display mod(2,2)
0
. display mod(3,2)
1
```

One useful device is a loop to get several results at once:

```
. forvalues i = 0/8 {
.     display "`i' " mod(`i', 3)
. }
```

Second, there is `generate`, typically followed by `list`:

```
. set obs 9
. generate mod3 = mod(_n - 1, 3)
. list mod3
```

You can use the observation numbers `_n`, which are integers 1 and up, to produce variables corresponding to successive integers.

Third, there is Mata, released in Stata 9:

```
. mata
: x = (0..8)
: mod(x, 3)
```

The first illustration, dividing 1, 2, and 3 by 2, points up a useful detail. Evidently, on division by 2, odd numbers have remainder 1 and even numbers, remainder 0. This example gives a way of characterizing odd and even in Stata. Suppose that you want to specify every other observation. Then

```
if mod(_n, 2) == 1
```

specifies odd-numbered observations and

```
if mod(_n, 2) == 0
```

specifies even-numbered observations. No new variable need be created, as Stata does the necessary calculations on the fly. We can be even more concise:

```
if mod(_n, 2)
```

selects odd observation numbers. Given mod(_n, 2), Stata evaluates it as 1 whenever _n is odd, which is nonzero and therefore true. Further,

```
if !mod(_n, 2)
```

selects even, as mod(_n, 2) is 0 whenever _n is even, but that result is flipped to 1 by the operator !, giving again 1, nonzero and true.

The idea extends easily to other divisors; for example, if mod(year, 10) == 0 or if !mod(year, 10) selects values of year divisible by 10 such as 1990 or 2000, and if !mod(year - 5, 10) selects years such as 1995 or 2005 (but not 1990 or 2000).

Now let us turn to sequences. For integers x from 0 up, mod(x, 3) is

0 1 2 0 1 2 0 1 2 ...

and for any positive integer y, mod(x, y) repeats cycles of 0 to $y - 1$. You may often want to add 1 to get, e.g.,

1 2 3 1 2 3 1 2 3 ...

Here you should use in Stata 1 + mod(x, 3) and in Mata 1 :+ mod(x, 3)—note the elementwise operator :+.

You could get such sequences in other ways. Using cond() (Kantor and Cox 2005), we could type for observation numbers _n that run 1 upwards cond(mod(_n, 3) == 0, 3, mod(_n, 3)), giving the same result.

Hence, you now have a basic recipe for generating repetitive sequences. You may know that this functionality is wired into **egen**'s seq() function, but the approach from first principles has merit, too.

Extracting digits is yet another application. In the shadow world between numbers and strings dwell numeric identifiers and run-together dates (20070328 for 28 March 2007) or times (112233 for 11:22:33). Whether such beasts are best processed as numbers or strings can be a close call. Conversion functions `real()` and `string()` are available to throw each to the other side of the divide.

Suppose that your beasts arrive as numeric. `mod(112233, 100)` extracts the last two digits. Hence, second arguments that are 10^k will extract the last k digits from integers.

Other subsequences of digits require a little more work. We could get the first two, the second two, and the third two digits like this:

```
. local first = floor(112233/10000)
. local second = floor(mod(112233, 10000) / 100)
. local third = mod(112233, 100)
. display 'first''second''third'
112233
```

For more on `floor()` and its twin `ceil()`, see Cox (2003). You could also use `int()` here. An alternative is to work with (say) `real(substr(string(112233),1,2))`.

Naturally, if what you are given is just 112233, you do not need Stata or even a computer to extract digits. Rather, these are examples of the kind you can try for yourself to see what is necessary to convert information given in variables from one form to another.

References

Biggs, N. L. 2002. *Discrete Mathematics*. Oxford: Oxford University Press.

Cox, N. J. 2003. Stata tip 2: Building with floors and ceilings. *Stata Journal* 3: 446–447.

Graham, R. L., D. E. Knuth, and O. Patashnik. 1994. *Concrete Mathematics: A Foundation for Computer Science*. Reading, MA: Addison–Wesley.

Kantor, D., and N. J. Cox. 2005. Depending on conditions: A tutorial on the cond() function. *Stata Journal* 5: 413–420.

Knuth, D. E. 1997. *The Art of Computer Programming. Volume 1: Fundamental Algorithms*. Reading, MA: Addison–Wesley.

The Stata Journal (2007)
7, Number 2, pp. 266–267

Stata tip 44: Get a handle on your sample

Ben Jann
ETH Zürich
Zürich, Switzerland
jann@soz.gess.ethz.ch

Researchers producing careful and reproducible statistical analyses need to keep track of precisely which observations are used by commands. Consider `regress` and similar commands as leading examples. The observations used will depend on any `if` or `in` conditions, any weights specified, and the incidence of missing values. Typically, you will want to look at results for `regress` together with those from other commands. For that you want the same observations to be used. Even when there is comparison with results for different subsets, you also need to monitor which observations are used by which commands.

`if` and `in` conditions and the use of weights are explicit in your command syntax, so you have only yourself to blame if you fail to consider their consequences. Stata, however, does not make a great fuss about excluding missing values from your analyses, so more attention is needed to this detail. Since most substantial statistical datasets contain missing values in at least some of the variables, the issue can arise often.

Researchers commonly start with a simple model and then add more predictors or covariates. At each step, some observations may be excluded because values are missing in the extra variables. As long as the proportion of missing values is not too large, you may not care much about them. However, correct interpretation of the results hinges on the subset of observations used remaining identical.

A brute force approach to the problem is to `keep` only those observations being used (or conversely to `drop` the others). But this method can create as many problems as it solves. Any number of different subsets analyzed would mean as many different datasets and consequent awkwardness in setting up comparisons.

A much better way to get a grip on the samples being used is to construct binary indicator or dummy variables that mark the observations used in any analysis. Their values should be 0 for excluded observations and 1 for included observations. With such variables, corresponding `if` conditions may be specified as desired.

Ado-file programmers (see [U] **17 Ado-files**) face a similar problem and have a special command to solve it. If you look at the source code of ado-files (using [P] **viewsource**; Cox 2006), you will often find the command `marksample touse` near the start and many `if 'touse'` qualifiers after. Although `marksample` can be used only within programs, [P] **mark** documents two other "programmer's commands", `mark` and `markout`, that prove to be handy outside ado-files, as I will now show.

Suppose that you are analyzing a dataset containing the variables x, y, and z, all of which contain some missing values; a group variable, g; and analytic weights, w. The analysis should be restricted to group `g == 1`. To ensure that the same observations are used throughout the analysis, type

```
. mark touse [aw=w] if g == 1
. markout touse x y z
```

at the beginning of your analysis and include `if touse` in all later commands, as in

```
. reg x y [aw=w] if touse
```

The first command, `mark`, generates a marker variable `touse` (read: "to use") that is set to 1 in observations satisfying the `if` qualifier and having a strictly positive, nonmissing weight and is set to 0 in all other observations.

The second command, `markout`, recodes `touse` to 0 if any of the specified variables contains missing. (If your data are `svyset`, you might want to omit the weights from the first command and add a third line reading `svymarkout touse`; see [SVY] **svymarkout**.)

There are other possible ways to generate the marker variable. You could, for example, use the `missing()` function (see [D] **functions**), or you could code

```
. quietly regress x y z [aw=w] if g==1
. generate byte touse = e(sample)
```

However, using `mark` and `markout` is simple and general. Often it is a good idea to `count if touse` and check that the number of observations used remains the same as that given.

If you need to have a one-liner, then define a program such as

```
program marktouse
version 8
syntax anything(id="markvar") [if] [in] [aw fw iw pw]
gettoken markvar varlist : anything
mark 'markvar' 'if' 'in' ['weight''exp']
markout 'markvar' 'varlist'
end
```

and use it as in

```
. marktouse touse x y z [pw=w] if g == 1
```

Reference

Cox, N. J. 2006. Stata tip 30: May the source be with you. *Stata Journal* 6: 149–150.

The Stata Journal (2007)
7, Number 2, pp. 268–271

Stata tip 45: Getting those data into shape

Christopher F. Baum
Department of Economics
Boston College
Chestnut Hill, MA 02467
baum@bc.edu

Nicholas J. Cox
Department of Geography
Durham University
Durham City, UK
n.j.cox@durham.ac.uk

Are your data in shape? That is, are they in the structure that you need to conduct the analysis you have in mind? Data sources often provide the data in a structure that is suitable for presentation but clumsy for statistical analysis. One of the key data management tools that Stata provides is reshape; see [D] **reshape**. If you need to modify the structure of your data, you should be familiar with reshape and its two functions: reshape wide and reshape long. In this tip, we discuss how two applications of reshape may be the solution to some knotty data management problems.

As a first example, consider this question posted on Statalist by an individual who has a dataset in the wide form:

country	tradeflow	Yr1990	Yr1991
Armenia	imports	105	120
Armenia	exports	90	100
Bolivia	imports	200	230
Bolivia	exports	80	115
Colombia	imports	100	105
Colombia	exports	70	71

He would like to reshape the data into long form:

country	year	imports	exports
Armenia	1990	105	90
Armenia	1991	120	100
Bolivia	1990	200	80
Bolivia	1991	230	115
Colombia	1990	100	70
Colombia	1991	105	71

We must exchange the roles of years and tradeflows in the original data to arrive at the desired structure, suitable for analysis as xt data. This exchange can be handled by two successive applications of reshape:

```
. reshape long Yr, i(country tradeflow)
(note: j = 1990 1991)
Data                                   wide    ->   long

Number of obs.                            6    ->      12
Number of variables                       4    ->       4
j variable (2 values)                           ->   _j
xij variables:
                            Yr1990 Yr1991       ->   Yr
```

This transformation swings the data into long form with each observation identified by country, tradeflow, and the new variable _j, taking on the values of year. We now perform reshape wide to make imports and exports into separate variables:

```
. rename _j year
. reshape wide Yr, i(country year) j(tradeflow) string
(note: j = exports imports)
Data                                   long    ->   wide

Number of obs.                           12    ->       6
Number of variables                       4    ->       4
j variable (2 values)             tradeflow    ->   (dropped)
xij variables:
                                   Yr           ->   Yrexports Yrimports
```

If we transform the data to wide form once again, the i() option contains country and year, as those are the desired identifiers on each observation of the target dataset. We specify that tradeflow is the j() variable for reshape, indicating that it is a string variable. The data now have the desired structure. Although we have illustrated this double-reshape transformation with only a few countries, years, and variables, the technique generalizes to any number of each.

As a second example of successive applications of reshape, consider the World Bank's World Development Indicators (WDI) dataset.[1] Their extract program generates a comma-separated value (CSV) database extract, readable by Excel or Stata, but the structure of those data hinders analysis as panel data. For a recent year, the header line of the CSV file is

```
"Series code","Country Code","Country Name","1960","1961","1962","1963",
"1964","1965","1966","1967","1968","1969","1970","1971","1972","1973",
"1974","1975","1976","1977","1978","1979","1980","1981","1982","1983",
"1984","1985","1986","1987","1988","1989","1990","1991","1992","1993",
"1994","1995","1996","1997","1998","1999","2000","2001","2002","2003","2004"
```

---

1. See http://econ.worldbank.org.

That is, each row of the CSV file contains a *variable* and *country* combination, with the columns representing the elements of the time series.[2]

Our target dataset structure is that appropriate for panel-data modeling, with the variables as columns and rows labeled by country and year. Two applications of `reshape` will again be needed to reach the target format. We first `insheet` (see [D] **insheet**) the data and transform the triliteral country code into a numeric code with the country codes as labels:

```
. insheet using wdiex.raw, comma names
. encode countrycode, generate(cc)
. drop countrycode
```

We then must address that the time-series variables are named `var4–var48`, as the header line provided invalid Stata variable names (numeric values) for those columns. We use `rename` (see [D] **rename**) to change v4 to d1960, v5 to d1961, and so on:

```
forv i=4/48 {
        rename v‘i’ d‘=1956+‘i’’
}
```

We now are ready to carry out the first `reshape`. We want to identify the rows of the reshaped dataset by both country code (`cc`) and `seriescode`, the variable name. The `reshape long` will transform a fragment of the WDI dataset containing two series and four countries:

```
. reshape long d, i(cc seriescode) j(year)
(note: j = 1960 1961 1962 1963 1964 1965 1966 1967 1968 1969 1970 1971 1972
> 1973 1974 1975 1976 1977 1978 1979 1980 1981 1982 1983 1984 1985 1986 1987
> 1988 1989 1990 1991 1992 1993 1994 1995 1996 1997 1998 1999 2000 2001 2002
> 2003 2004)
```

Data	wide	->	long
Number of obs.	7	->	315
Number of variables	48	->	5
j variable (45 values)		->	year
xij variables:			
d1960 d1961 ... d2004		->	d

---

2. A variation occasionally encountered will resemble this structure, but with periods in reverse chronological order. The solution here can be used to deal with that problem as well.

```
. list in 1/15
```

	cc	seriesc~e	year	countryname	d
1.	AFG	adjnetsav	1960	Afghanistan	.
2.	AFG	adjnetsav	1961	Afghanistan	.
3.	AFG	adjnetsav	1962	Afghanistan	.
4.	AFG	adjnetsav	1963	Afghanistan	.
5.	AFG	adjnetsav	1964	Afghanistan	.
6.	AFG	adjnetsav	1965	Afghanistan	.
7.	AFG	adjnetsav	1966	Afghanistan	.
8.	AFG	adjnetsav	1967	Afghanistan	.
9.	AFG	adjnetsav	1968	Afghanistan	.
10.	AFG	adjnetsav	1969	Afghanistan	.
11.	AFG	adjnetsav	1970	Afghanistan	-2.97129
12.	AFG	adjnetsav	1971	Afghanistan	-5.54518
13.	AFG	adjnetsav	1972	Afghanistan	-2.40726
14.	AFG	adjnetsav	1973	Afghanistan	-.188281
15.	AFG	adjnetsav	1974	Afghanistan	1.39753

The rows of the data are now labeled by year, but one problem remains: all variables for a given country are stacked vertically. To unstack the variables and put them in shape for xtreg (see [XT] **xtreg**), we must carry out a second **reshape** that spreads the variables across the columns, specifying cc and year as the i variables and seriescode as the j variable. Since that variable has string content, we use the **string** option.

```
. reshape wide d, i(cc year) j(seriescode) string
(note: j = adjnetsav adjsavC02)
```

Data	long	->	wide
Number of obs.	315	->	180
Number of variables	5	->	5
j variable (2 values)	seriescode	->	(dropped)
xij variables:			
	d	->	dadjnetsav dadjsavC02

```
. order cc countryname

. tsset cc year
        panel variable:  cc (strongly balanced)
        time variable:   year, 1960 to 2004
```

After this transformation, the data are now in shape for xt modeling, tabulation, or graphics.

As illustrated here, the **reshape** command can transform even the most inconvenient data structure into the structure needed for your research. It may take more than one application of **reshape** to get there from here, but it can do the job.

The Stata Journal (2007)
7, Number 2, pp. 272–274

Stata tip 46: Step we gaily, on we go

Richard Williams
University of Notre Dame
Notre Dame, IN 46556
richard.a.williams.5@nd.edu

The `nestreg` and `stepwise` prefix commands allow users to estimate sequences of nested models. With `nestreg`, you specify the order in which variables are added to the model. So, for example, a first model might include only demographic characteristics of subjects, a second could add attitudinal measures, and a third could add interaction terms. Conversely, with `stepwise`, the order in which variables enter the model is determined empirically. With forward selection, the variable or block of variables that most improves fit will be entered first, followed by the variable or variables that most improve fit given the variables already in the model, and so forth. Variables that do not meet some specified level of significance will never enter the model.

Despite their similarities, the two commands differ dramatically in the amount of detail that they provide. `stepwise` gives the estimates for the final model it fits but tells little about the intermediate models other than the order in which variables were entered. `nestreg`, on the other hand, offers a wealth of information. The results from each intermediate model can be printed and their estimates stored for later use. Particularly useful is that `nestreg` offers several measures of contribution to model fit. Wald statistics, likelihood-ratio chi-squareds, R^2 and change-in-R^2 statistics, and Bayesian information criterion (BIC) and Akaike information criterion (AIC) measures are available for each intermediate model. Such measures provide a variety of ways of assessing the importance and effect of each variable or set of variables added to the model.

When forward selection is used, there is a relatively easy way to make the results from `stepwise` as informative and detailed as those provided by `nestreg`. Simply fit the models with `stepwise`, and then refit the models with `nestreg`, listing variables in the order they were added by `stepwise`. For example,

```
. sysuse auto
(1978 Automobile Data)

. stepwise, pe(.05): regress price mpg weight length foreign
                    begin with empty model
p = 0.0000 <  0.0500  adding  weight
p = 0.0000 <  0.0500  adding  foreign
p = 0.0069 <  0.0500  adding  length
```

Source	SS	df	MS
Model	348565467	3	116188489
Residual	286499930	70	4092856.14
Total	635065396	73	8699525.97

```
Number of obs =      74
F(  3,    70) =   28.39
Prob > F      =  0.0000
R-squared     =  0.5489
Adj R-squared =  0.5295
Root MSE      =  2023.1
```

price	Coef.	Std. Err.	t	P>|t|	[95% Conf.	Interval]
weight	5.774712	.9594168	6.02	0.000	3.861215	7.688208
foreign	3573.092	639.328	5.59	0.000	2297.992	4848.191
length	-91.37083	32.82833	-2.78	0.007	-156.8449	-25.89679
_cons	4838.021	3742.01	1.29	0.200	-2625.183	12301.22

```
. nestreg, quietly store(m): regress price weight foreign length if e(sample)
Block  1: weight
Block  2: foreign
Block  3: length
```

Block	F	Block df	Residual df	Pr > F	R2	Change in R2
1	29.42	1	72	0.0000	0.2901	
2	29.59	1	71	0.0000	0.4989	0.2088
3	7.75	1	70	0.0069	0.5489	0.0499

Specifying the quietly option omitted the estimates from the intermediate models from the output while still showing the various model change statistics. You may wish to omit quietly if, for example, changes in coefficients across models as additional variables are added are of interest. The store(m) option stored the estimates from the three models as m1, m2, and m3. Storing the results can be useful if we want to replay the results, format the output with some other program, or do more comparisons across models, say, model m1 versus model m3. Using if e(sample) guarantees that the same observations are being analyzed by both nestreg and stepwise. With stepwise, observations with missing data on any of the variables specified get excluded from the analysis, even if those variables do not enter the final model. Most critically, the block residual statistics reported by nestreg show us how much the addition of each variable increased R^2 and the statistical significance of that change. This approach provides a much more tangible feel for the importance and contribution of each variable than does stepwise alone.

If we would also like to see likelihood-ratio contrasts between models, as well as the BIC and AIC statistics for each intermediate model, just add the lr option:

```
. nestreg, quietly lr: regress price weight foreign length if e(sample)
Block  1: weight
Block  2: foreign
Block  3: length
```

Block	LL	LR	df	Pr > LR	AIC	BIC
1	-683.0354	25.35	1	0.0000	1370.071	1374.679
2	-670.1448	25.78	1	0.0000	1346.29	1353.202
3	-666.2613	7.77	1	0.0053	1340.523	1349.739

Naturally, keep the usual cautions concerning stepwise procedures in mind. For example, because multiple tests are being conducted, the reported p-values are inaccurate. The researcher may therefore wish to use a more stringent significance level for variable entry, e.g., .01, or use a Bonferroni or other adjustment. Chance alone could cause some variables to enter the model, and a different sample might produce a different final model. Forward and backward selection procedures can also result in different final models. But if you are clear that stepwise selection is appropriate and is being conducted correctly, then combining `stepwise` and `nestreg` should be helpful.

The Stata Journal (2007)
7, Number 2, pp. 275–279

Stata tip 47: Quantile–quantile plots without programming

Nicholas J. Cox
Durham University
Durham City, UK
n.j.cox@durham.ac.uk

Quantile–quantile (Q–Q) plots are one of the staples of statistical graphics. Wilk and Gnanadesikan (1968) gave a detailed and stimulating review that still merits close reading. Cleveland (1993, 1994) gave more recent introductions. Here I look at their use for examining fit to distributions. The quantiles observed for a variable, which are just the data ordered from smallest to largest, may be plotted against the corresponding quantiles from some theoretical distribution. A good fit would yield a simple linear pattern. Marked deviations from linearity may indicate characteristics such as skewness, tail weight, multimodality, granularity, or outliers that do not match those of the theoretical distribution. Many consider such plots more informative than individual figures of merit or hypothesis tests and feature them prominently in intermediate or advanced surveys (e.g., Rice 2007; Davison 2003).

Official Stata includes commands for plots of observed versus expected quantiles for the normal (`qnorm`) and chi-squared (`qchi`) distributions. User-written commands can be found for other distributions. You might guess that such graphics depend on the provision of dedicated programs, but much can be done interactively just by combining some basic commands. Indeed, you can easily experiment with variations on the standard plots not yet provided in any Stata program.

Statistical and Stata tradition dictate that we start with the normal distribution and the `auto` dataset. In a departure from tradition, generate `gpm` (gallons per 100 miles) as a reciprocal of `mpg` (miles per gallon) scaled to convenient units and examine its fit to normality. You can calculate the ranks and sample size by using `egen`:

```
. use http://www.stata-press.com/data/r9/auto
(1978 Automobile Data)
. gen gpm = 100 / mpg
. label var gpm "gallons / 100 miles"
. egen rank = rank(gpm)
. egen n = count(gpm)
```

These `egen` functions handle any missing values automatically and can easily be combined with any extra `if` and `in` conditions. You may like to specify the `unique` option with `rank()` if you have many ties on your variable. If you want to fit separate distributions to distinct groups, apply the `by:` prefix, say,

```
. by foreign, sort: egen rank = rank(gpm)
. by foreign: egen n = count(gpm)
```

Next choose a formula for plotting positions given rank i and count n. These positions are cumulative probabilities associated with the data. The formula i/n would imply that no value could be larger than the largest observed in the sample and would render the normal quantile unplottable for the same extreme. The formula $(i-1)/n$ would be similarly objectionable at the opposite extreme. Various alternatives have been proposed, typically $(i-a)/(n-2a+1)$ for some a: you may choose for yourself. qnorm has $i/(n+1)$ (i.e., $a=0$) wired in, but let us take $a=0.5$ to emphasize our freedom. A minimal plot is now within reach using invnormal(), the normal quantile or inverse cumulative distribution function. Figure 1 is our first stab, with separate fits for the two groups of cars.

```
. gen pp = (rank - 0.5) / n
. gen normal = invnormal(pp)
. scatter gpm normal if foreign, ms(oh) ||
> scatter gpm normal if !foreign, ms(S) yla(, ang(h))
> legend(order(1 "Foreign" 2 "Domestic") ring(0) pos(5) col(1))
> xti(standard normal)
```

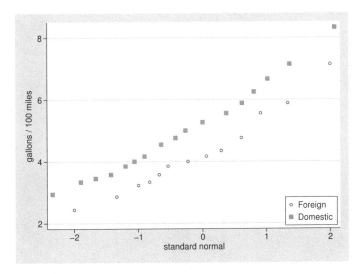

Figure 1: Normal probability plots for gallons per 100 miles for foreign and domestic cars

You might want to fit means and standard deviations explicitly. The easiest way is once again to use egen:

```
. by foreign: egen mean = mean(gpm)
. by foreign: egen sd = sd(gpm)
. gen normal2 = mean + sd * normal
. scatter gpm normal2, by(foreign, note("") legend(off)) ||
> function equality = x, ra(normal2) yla(, ang(h))
> xti(fitted normal) yti(gallons / 100 miles)
    Graph not shown to save space
```

Already with just a few lines we can do something not available with `qnorm`: plotting two or more groups. We can superimpose, as in figure 1, or juxtapose, as in the last example.

Variants of the basic Q–Q plot are also close at hand. Wilk and Gnanadesikan (1968) suggested some possibilities. As is standard practice in examining model fit, we may subtract the general tilt of the Q–Q plot by looking at the residuals, the differences between observed and expected quantiles. These may be plotted against either the expected quantiles or the plotting positions. The two graphs convey similar information. These difference quantile plots might be called DQ plots for short. DQ plots are in essence more demanding than standard Q–Q plots, as they make discrepancies from expectation more evident. As with residual plots, the reference line is no longer a diagonal line of equality but rather the horizontal line of zero difference or residual. Figure 2 shows the two possibilities mentioned. Although `gpm` is more nearly Gaussian than `mpg`, some marked skewness remains. Lowess or other smoothing could be used to identify any systematic structure.

```
. gen residual = gpm - normal2

. scatter residual normal2 if foreign, ms(oh) ||
> scatter residual normal2 if !foreign, ms(S)
> legend(order(1 "Foreign" 2 "Domestic") pos(5) ring(0) col(1))
> yla(, ang(h)) yli(0) xti(fitted normal) saving(graph1)
(file graph1.gph saved)

. scatter residual pp if foreign, ms(oh) ||
> scatter residual pp if !foreign, ms(S)
> legend(order(1 "Foreign" 2 "Domestic") pos(5) ring(0) col(1))
> yla(, ang(h)) yli(0) xti(plotting position) saving(graph2)
(file graph2.gph saved)

. graph combine graph1.gph graph2.gph
```

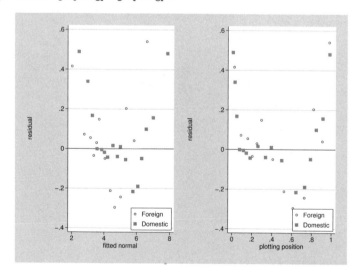

Figure 2: DQ plots for gallons per 100 miles for foreign and domestic cars and normal distribution. Residual versus (left) fitted quantile and (right) plotting position.

The example distribution, the normal, is specified by a location parameter and a scale parameter. This fact gives the flexibility of either fitting parameters or not fitting parameters first. If the theoretical distribution is also specified by one or more shape parameters, we would need to specify those first.

Turning away from the normal, we close with different examples. Q–Q plots and various relatives are prominent in work on the statistics of extremes (e.g., Coles 2001; Reiss and Thomas 2001; Beirlant et al. 2004) and more generally in work with heavy- or fat-tailed distributions. One way of using Q–Q plots is as an initial exploratory device, comparing a distribution, or its more interesting tail, with some reference distribution. For exponential distributions,

```
. generate exponential = -ln(1 - pp)
```

and plot data against that. On such plots, distributions heavier tailed than the exponential will be convex down and those lighter tailed will be convex up (Beirlant et al. 2004). For work with maximums, the Gumbel distribution is a basic starting point.

```
. generate Gumbel = -ln(-ln(pp))
```

Figure 3 is a basic Gumbel plot for annual maximum sea levels at Port Pirie in Australia (data for 1923–1987 from Coles 2001).

```
. scatter level Gumbel, yla(, ang(h)) xti(standard Gumbel)
```

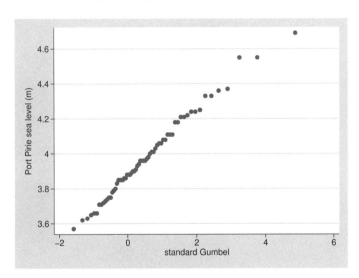

Figure 3: Basic Gumbel plot for annual maximum sea levels at Port Pirie in Australia

The generally good linearity encourages a more formal fit. Convex or concave curves would have pointed to fitting other members of the generalized extreme value distribution family.

References

Beirlant, J., Y. Goegebeur, J. Segers, and J. Teugels. 2004. *Statistics of Extremes: Theory and Applications*. New York: Wiley.

Cleveland, W. S. 1993. *Visualizing Data*. Summit, NJ: Hobart Press.

———. 1994. *The Elements of Graphing Data*. Summit, NJ: Hobart Press.

Coles, S. 2001. *An Introduction to Statistical Modeling of Extreme Values*. London: Springer.

Davison, A. C. 2003. *Statistical Models*. Cambridge: Cambridge University Press.

Reiss, R.-D., and M. Thomas. 2001. *Statistical Analysis of Extreme Values with Applications to Insurance, Finance, Hydrology, and Other Fields*. Basel: Birkhäuser.

Rice, J. A. 2007. *Mathematical Statistics and Data Analysis*. Belmont, CA: Duxbury.

Wilk, M. B., and R. Gnanadesikan. 1968. Probability plotting methods for the analysis of data. *Biometrika* 55: 1–17.

The Stata Journal (2007)
7, Number 3, pp. 434–435

Stata tip 48: Discrete uses for uniform()[1]

Maarten L. Buis
Department of Social Research Methodology
Vrije Universiteit Amsterdam
Amsterdam, The Netherlands
m.buis@fsw.vu.nl

The `uniform()` function produces random draws from a uniform distribution between 0 and 1 ([D] **functions**). `uniform()` is an unusual function. It takes no arguments, although the parentheses are essential to distinguish it from a variable name, and it returns different values each time it is invoked—as many as are needed. Thus, if `uniform()` is used to generate a variable, a different value is created in each observation. This Stata tip focuses on one of its many uses: creating random draws from a discrete distribution where each possible value has a known probability.

A uniform distribution between 0 and 1 means that each number between 0 and 1 is equally likely. So the probability that a random draw from a uniform distribution has a value less than 0.5 is 50%, the probability that such a random draw has a value less than 0.6 is 60%, and so on. The first example below shows how this fact can be used to create a random variable, where the probability of drawing 1 is 60% and that of drawing 0 is 40%. (Kantor and Cox [2005] give a tutorial on the `cond()` function.)

```
gen draw = cond(uniform() < .6, 1, 0)
```

The same result can be achieved even more concisely, given that in Stata a true condition is evaluated as 1 and a false condition as 0 (Cox 2005). `uniform() < .6` is true, and thus evaluated as 1, whenever the function produces a random number less than .6 and is false, and thus evaluated as 0, whenever that is not so.

```
gen draw = uniform() < .6
```

The probability need not be constant. Suppose that the probability of drawing 1 depends on a variable x. We can simulate data for a logistic regression with a constant of -1 and an effect of x of 1. In this example, the variable x contains draws from a standard normal distribution.

```
gen x = invnorm(uniform())
gen draw = uniform() < invlogit(-1 + x)
```

Nor is this method limited to random variables with only two values. Consider a distribution in which 1 has probability 30%, 2 probability 45%, and 3 probability 25%.

```
gen rand = uniform()
gen draw = cond(rand < .3,  1, cond(rand < .75, 2, 3))
```

---

1. Editors' note: As of Stata 10.1, the function `uniform()` has been deprecated in favor of `runiform()`. Random draws from a normal distribution are available via the function `rnormal()`. See [D] **functions** or `help random number functions`.

A special case of this distribution is one for k integers, say, 1 to k, in which each value is equally likely. Suppose that we simulate throwing a six-sided die, so that values from 1 to 6 are assumed to have probability 1/6. So we need to map all values up to 1/6 to 1, those up to 2/6 to 2, and so forth. That goal is easily achieved by multiplying by 6 and rounding up, using the ceiling function `ceil()`. Other applications of `ceil()` were discussed in Stata tip 2 (Cox 2003).

```
gen draw = ceil(6 * uniform())
```

We can use the same principle to simulate draws from a binomial distribution. Recall that a binomial distribution with parameters n and p is the distribution of the number of "successes" out of n trials when the probability of success in each trial is p. One way of sampling from this distribution is to do just that; i.e., draw n numbers from a uniform distribution, declare each number a success if it is less than p, and then count the number of successes (Devroye 1986, 524). Here Mata is convenient. Its equivalent to `uniform()`, `uniform(r, c)`, creates an $r \times c$ matrix filled with random draws from the uniform distribution. Thus we can create a new variable, `draw`, containing draws from a binomial(100, .3) distribution:

```
mata:
n = 100
p = .3
draw = J(st_nobs(),1,.)              // matrix to store results
for(i = 1; i <= rows(draw); i++) {   // loop over observations
    trials = uniform(1, n)           // create n trials
    successes = trials :< p          // success = 1 failure = 0
    draw[i,1] = rowsum(successes)    // count the successes
}
idx = st_addvar("int", "draw")
st_store(., idx, draw)               // store the variable
end
```

This code is deliberately spun to make its logit clear. Mata learners and experts alike might enjoy working out how to eliminate the loop and how to use fewer variables, while also pondering the possibility of a problem with memory demand for large datasets.

References

Cox, N. J. 2003. Stata tip 2: Building with floors and ceilings. *Stata Journal* 3: 446–447.

———. 2005. FAQ: What is true or false in Stata?
    http://www.stata.com/support/faqs/data/trueorfalse.html.

Devroye, L. 1986. *Non-Uniform Random Variate Generation*. New York: Springer.

Kantor, D., and N. J. Cox. 2005. Depending on conditions: A tutorial on the cond() function. *Stata Journal* 5: 413–420.

The Stata Journal (2007)
7, Number 3, pp. 436–437

Stata tip 49: Range frame plots

Scott Merryman
Risk Management Agency
Kansas City, MO
scott.merryman@gmail.com

One of Edward Tufte's principles for good graph design is to erase nondata ink (Tufte 2001). The axes on a typical scatterplot will extend beyond the range of the observed data. You can minimize the nondata ink by having the axis lines extend only over the range of the data. This can be accomplished by not having the lines extend through the plot region and the region's margin:

```
. sysuse auto
. twoway scatter mpg price, ylabel(minmax, nogrid) xlabel(minmax)
> yscale(nofextend) xscale(nofextend) plotregion(margin(5 2 5 2))
```

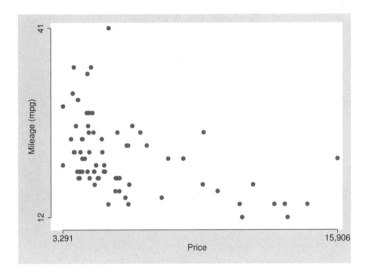

`nofextend` is an undocumented axis-scale option.

To position the origin of the scatterplot at $(0, 10)$, you can adjust the `plotregion(margin())` by the percentage of the observed axis to the total axis length.

(*Continued on next page*)

```
. sysuse auto
. summarize price, meanonly
. local w = (1 - (r(max) - r(min))/(r(max) - 0))*100
. summarize mpg, meanonly
. local h = (1 - (r(max) - r(min))/(r(max) - 10) )*100
. twoway scatter mpg price, ylabel(minmax, nogrid) xlabel(minmax)
> yscale(nofextend) xscale(nofextend) plotregion(margin(`w' 2 `h' 2))
```

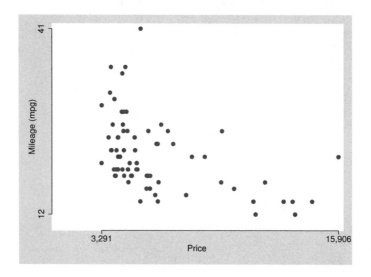

Reference

Tufte, E. R. 2001. *The Visual Display of Quantitative Information.* 2nd ed. Cheshire, CT: Graphics Press.

The Stata Journal (2007)
7, Number 3, pp. 438–439

Stata tip 50: Efficient use of summarize

Nicholas J. Cox
Department of Geography
Durham University
Durham City, UK
n.j.cox@durham.ac.uk

The `summarize` command must be one of the most commonly used Stata commands. Yet, strangely, one of its options is often not used, even though it can be the best solution to a user's problem. Here I flag this neglected `meanonly` option and speculate briefly on why it is often overlooked.

If you fire up `summarize, meanonly`, no results appear in the Results window for you to examine. This lack is deliberate. The option leaves r-class results in memory. (If you are unclear what that means, start with the online help for `return`.) The user must access those results by typing `return list` to see what they are or by feeding one or more results to something else, such as an explicit `display`, `generate`, or `replace` statement. Accessing the saved results should be done promptly after the `summarize, meanonly` command has finished, because results are ephemeral and will not survive beyond the next r-class command that is issued.

The `meanonly` option leaves in memory the mean, as the name implies, in $r(mean)$. However, contrary to what you might guess, it also leaves behind the count of nonmissing values, the sum, the weighted sum, the minimum, and the maximum in appropriately named results. These results are for the last-named variable. Thus, although invoking `summarize, meanonly` with two or more variables is legal, doing so is utterly pointless because results for all but the last will disappear and machine time will be wasted.

Incidentally, if all you want is a count, the `count` command offers a more direct solution; see [D] **count** and Cox (2007).

The difference between `summarize, meanonly` and `summarize` with no options is that the latter also calculates the variance and its square root, the standard deviation. The reason for the `meanonly` option is that this last calculation can be fairly time consuming in large datasets. Thus, if you need to use only one or more of the results left behind after `summarize, meanonly`, then specifying the option will be sensible. Programs or do-files that will be used repeatedly and/or on large datasets are especially suitable. Budding programmers can entertain themselves by identifying StataCorp programs that passed up opportunities for using `summarize, meanonly`. This issue underscores an old joke that you can always speed up a program that was originally written to run slowly.

As a concrete example, one common task is cycling over a set of categories defined by one or more variables. An easy way to do this is to use `egen, group()` to create a variable with integer values 1 and up (and, optionally, value labels with informative text). When you do not know the number of categories in advance,

```
. summarize group, meanonly
```

produces the maximum of `group`, which is the same as the number of categories present. Thus we can feed `r(max)` to whatever code that needs it, possibly a `forvalues` loop.

A small problem remains of explaining why people often overlook this `meanonly` option. I have three guesses. First, `summarize` is one of the commands that Stata users learn early. Typically, it quickly becomes clear that `summarize` does various things and `summarize, detail` does even more. Thus, users tend to feel that they are familiar with the command and do not study its help carefully. Second, the name `meanonly` is in some ways unfortunate and misleading, because much more than the mean is produced. Perhaps a synonym such as `summarize, short` would be a good idea. (Dropping the `meanonly` name is not likely, given the number of programs and commands that would break.) Third, the explanation of `meanonly` in the manual at [R] **summarize** does not give the complete picture on this option.

Reference

Cox, N. J. 2007. Speaking Stata: Making it count. *Stata Journal* 7: 117–130.

The Stata Journal (2007)
7, Number 3, pp. 440–443

Stata tip 51: Events in intervals

Nicholas J. Cox
Department of Geography
Durham University
Durham City, UK
n.j.cox@durham.ac.uk

Observations in panel or longitudinal datasets are often for irregularly spaced times. Patients may arrive for consultation or treatment, or sites may be visited in the field, at arbitrary times, or other human or natural events may occur with unpredictably uneven gaps. Geophysicists might record earthquakes or political scientists might record incidents of unrest; in either case, events occur with their own irregularity. Such examples could be multiplied. One way researchers seek structure in such data is by counting or summarizing data for each panel over chosen time windows. Usually we look backward: How many times did something happen in the previous 6 months? What was the average of some important variable over observations in the previous 30 days?

To see the precise problem in Stata terms, consider better behaved data in which we have regular observations, say, monthly or daily. Then we can use windows with fixed numbers of observations to calculate the summaries required. `rolling` ([TS] **rolling**) can be especially useful here. This scenario suggests one solution: a dataset with irregular data can be made regular by inserting observations for dates not present in the data. `tsfill` ([TS] **tsfill**) is the key command. In turn the downside of that solution is evident: the bulked-out dataset could be many times larger, even though it carries no extra information.

A more direct solution is possible, typically requiring a few lines of Stata code. Once you grasp the solution, modifying the code for similar problems is easy.

Suppose first that you want to count certain kinds of observations, say, how many times something happened in the previous 30 days. We assume that the data include an identifier (say, `id`) and a daily date (say, `date`) among other variables. A good technique to consider is using the `count` command (Cox 2007a). First, initialize a count variable. Our example will count observations with high blood pressure, so the variable name reflects that:

```
gen n_high_bp = .
```

The idea is to loop over the observations, looking at each one in turn. A basic count will be

```
count if some condition is true &
         observation is in the same panel as this one &
         time is within interval of interest relative to this one
```

The example above is part Stata code, part pseudocode. The parts in *slanted type* are pseudocode. `count` will produce a number in your Results window, but that is less

important than `count`'s leaving the result in `r(N)`. We must grab that result before
something else overwrites it or it just disappears. We can grab the result and use it:

```
replace n_high_bp = r(N) in this observation
```

We want to repeat this step for each observation. You may know that you can use
`forvalues`, often abbreviated `forval`, for automating a loop easily (see Cox 2002 for a
tutorial). Suppose that you have 4,567 observations. Then you can type

```
forval i = 1/4567 {
        count if conditions are all satisfied
        replace n_high_bp = r(N) in 'i'
}
```

Naturally, your having 4,567 observations is unlikely. So, you could just substitute
the correct number for 4,567, or you could think more generally. _N is the number of
observations.

```
local N = _N
forval i = 1/'N' {
        count if conditions are all satisfied
        replace nhighbp = r(N) in 'i'
}
```

`forval` is fussy in its syntax, so we cannot use _N directly. The `local` statement
sets a local macro, N, to contain its value. Once that exists, we can use its contents
by referring to 'N'. As you might guess, i and 'i' refer to another local macro, which
the `forvalues` loop brings into being. Each time around the loop it takes on values
between 1 and the number of observations.

There are three slots to fill in the pseudocode. Here are three examples to match:

> *some condition is true*
> ```
> inrange(sys_bp, 120, .)
> ```
> *observation is in the same panel as this one*
> ```
> id == id['i']
> ```
> *time is within interval of interest relative to this one*
> ```
> inrange(date['i'] - date, 1, 30)
> ```

Our examples use `inrange()` twice. In the first, we suppose that we are counting
how often systolic blood pressure was 120 mm Hg or higher. The `inrange()` condition
here has one special and useful feature: it excludes missing values. That is, for example,
`inrange(., 120, .)` is 0 (false). I do not expect you to find this behavior intuitive, but
it is a feature. In the second, the previous 30 days is specified. For more on `inrange()`,
see Cox (2006).

Now we can put it all together.

```
gen n_high_bp = .
local N = _N
quietly forval i = 1/'N' {
        count if inrange(sys_bp, 120, .) & ///
                id == id['i']            & ///
                inrange(date['i'] - date, 1, 30)
        replace n_high_bp = r(N) in 'i'
}
```

A new detail here is the `quietly` added to the loop to stop a long list of results from being shown. Doing so is not essential. Indeed, at a debugging stage, seeing a stream of output, and being able to check that the results are as desired, is useful and reassuring.

Experienced programmers usually reduce that by one line, starting like this:

```
gen n_high_bp = .
quietly forval i = 1/'= _N' {
```

The shortcut here is documented under the help for `macro`. We are evaluating an expression, here just _N, and using its result, all within the space of the command line.

Stata users sometimes want to do something like this:

```
quietly forval i = 1/'= _N' {
        count if inrange(sys_bp, 120, .) & ///
                id == id['i']            & ///
                inrange(date['i'] - date, 1, 30)
        gen n_high_bp = r(N) in 'i'
}
```

That code will fail the second time around the loop. The first time around the loop, when i is 1, all will be fine. The new variable `n_high_bp` will be generated. `r(N)` will be put into `n_high_bp[1]`. All the other values of `n_high_bp` will be born as missing. However, the second time around the loop, when i is 2, the `generate` command is illegal, as the `n_high_bp` variable already exists, and you cannot `generate` it again.

The consequence is that within the loop we need to use `replace`. In turn, we need to initialize the variable outside and before the loop (because, conversely, you cannot `replace` something that does not yet exist). Initializing it to missing is good practice, even when we know that the program will overwrite the value in each observation.

There are some disadvantages to this approach. Mainly, it will be a bit slow, especially with large datasets. Having to spell out a few lines of code every time you do something similar could also prove tedious. That task could be an incentive to wrap up the code in a do-file or even a program.

More positively, the logic here should seem straightforward and transparent and fairly easy to modify for similar problems. The key will usually be to pick up whatever we need as a saved result. Suppose that we want to record the mean systolic blood pressure over measurements in the last 30 days. The main change is the use of the `summarize` command rather than the `count` command.

```
gen mean_sys_bp = .
quietly forval i = 1/'= _N' {
    summarize sys_bp if id == id['i'] & ///
                      inrange(date['i'] - date, 1, 30), meanonly
    replace mean_sys_bp = r(mean) in 'i'
}
```

For the `meanonly` option of `summarize` and its advantages, see the previous Stata tip (Cox 2007b).

Naturally, there are occasional problems in which the condition that we are considering only observations in the same panel is inappropriate. For those problems, remove or change code like `id == id['i']`.

Finally, the technique is readily adaptable to other kinds of windows, say, with regard to intervals of any predictor or controlling variable.

References

Cox, N. J. 2002. Speaking Stata: How to face lists with fortitude. *Stata Journal* 2: 202–222.

———. 2006. Stata tip 39: In a list or out? In a range or out? *Stata Journal* 6: 593–595.

———. 2007a. Speaking Stata: Making it count. *Stata Journal* 7: 117–130.

———. 2007b. Stata tip 50: Efficient use of summarize. *Stata Journal* 7: 438–439.

The Stata Journal (2007)
7, Number 4, pp. 582–583

Stata tip 52: Generating composite categorical variables

Nicholas J. Cox
Department of Geography
Durham University
Durham City, UK
n.j.cox@durham.ac.uk

If you have two or more categorical variables, you may want to create one composite categorical variable that can take on all the possible joint values. The canonical example for Stata users is given by cross-combinations of `foreign` and `rep78` in the `auto` data. Setting aside missings, `foreign` takes on values of 0 and 1, and `rep78` takes on values of 1, 2, 3, 4, and 5. Hence there are ten possible joint values, which could be 0 and 1, 0 and 2, and so forth. As it happens, only eight occur in the data. If we add the value labels attached to `foreign`, we have Domestic 1, Domestic 2, and so forth.

Writing the values like that raises the question of whether these cross-combinations will be better expressed as string variables or as numeric variables with value labels. On the whole, an integer-valued numeric variable with value labels defined and attached is the best arrangement for any categorical variable, but a string variable may also be convenient, especially if you are producing a kind of composite identifier.

A method often seen is to produce string variables with `tostring` (see [D] **destring**), for example,

```
. tostring foreign rep78, generate(Foreign Rep78)
. gen both = Foreign + Rep78
```

Naturally, there are endless minor variations on this method. A small but useful improvement is to insert a space or other punctuation:

```
. gen both = Foreign + " " + Rep78
```

However, this method is not especially good. `tostring` is really for correcting mistakes, whether attributable to human fault or to some software used before you entered Stata: some variable that should be string is in fact numeric. You need to correct that mistake. `tostring` is a safe way of doing that.

That intended purpose does not stop `tostring` being useful for things for which it was not intended, but there are two specific disadvantages to this method:

1. This method needs two lines, and you can do it in one. That is a little deal.

2. This method could lose information, especially for variables with value labels or with noninteger values. That is, potentially, a big deal.

The second point may suggest using `decode` instead, but my suggestions differ. A better method is to use `egen, group()`. See [D] **egen**.

```
. egen both = group(foreign rep78), label
```

This command produces a new numeric variable, with integer values 1 and above, and value labels defined and attached. Particularly, note the `label` option, which is frequently overlooked.

This method has several advantages:

1. One line.

2. No loss of information. Observations that are identical on the arguments are identical on the results. Value labels are used, not ignored. Distinct noninteger values will also remain distinct.

3. The label is useful—indeed essential—for tables and graphs to make sense.

4. Efficient storage.

5. Extends readily to three or more variables.

Another fairly good method is to use `egen, concat()`.

```
. egen both = concat(foreign rep78), decode p(" ")
```

This command creates a string variable, so it is less efficient for data storage and is less versatile for graphics or modeling. Compared with `tostring`, the advantages are

1. One line.

2. You can mix numeric and string arguments. `concat()` will calculate what is needed.

3. You can use the `decode` option to use value labels on the fly.

4. You can specify punctuation as separator, here a blank.

5. Extends to three or more variables.

The Stata Journal (2007)
7, Number 4, pp. 584–586

Stata tip 53: Where did my p-values go?

Maarten L. Buis
Department of Social Research Methodology
Vrije Universiteit Amsterdam
Amsterdam, The Netherlands
m.buis@fsw.vu.nl

A useful item in the Stata toolkit is the returned result. For example, after most estimation commands, parameter estimates are stored in a matrix `e(b)`. However, these commands do not return the t statistics, p-values, and confidence intervals for those parameter estimates. The aim here is to show how to recover those statistics by using the results that are returned. Consider the following OLS regression:

```
. sysuse auto
(1978 Automobile Data)

. regress price mpg foreign

      Source |       SS       df       MS              Number of obs =      74
-------------+------------------------------           F(  2,    71) =   14.07
       Model |  180261702      2  90130850.8           Prob > F      =  0.0000
    Residual |  454803695     71  6405685.84           R-squared     =  0.2838
-------------+------------------------------           Adj R-squared =  0.2637
       Total |  635065396     73  8699525.97           Root MSE      =  2530.9

------------------------------------------------------------------------------
       price |      Coef.   Std. Err.      t    P>|t|     [95% Conf. Interval]
-------------+----------------------------------------------------------------
         mpg |  -294.1955   55.69172    -5.28   0.000    -405.2417   -183.1494
     foreign |   1767.292    700.158     2.52   0.014     371.2169    3163.368
       _cons |   11905.42   1158.634    10.28   0.000     9595.164    14215.67
------------------------------------------------------------------------------
```

1   t statistic

The t statistic can be calculated from $t = (\widehat{b} - b)/\text{se}$, where $\widehat{b}$ is the estimated parameter, b is the parameter value under the null hypothesis, and se is the standard error. The null hypothesis is usually that the parameter equals zero; thus we have $t = \widehat{b}/\text{se}$. The t statistic for one parameter (`foreign`) can be calculated by

```
. di _b[foreign]/_se[foreign]
2.5241336
```

All the parameter estimates are also returned in the matrix `e(b)`. A vector of all standard errors is a bit harder to obtain; they are the square roots of the diagonal elements of the matrix `e(V)`. In Mata that vector can be created by typing `diagonal(cholesky(diag(V)))`. Continuing the example, a vector of all t statistics can be computed within Mata by

```
: b = st_matrix("e(b)")'
: V = st_matrix("e(V)")
```

```
: se = diagonal(cholesky(diag(V)))
: b :/ se
                    1

    1 │    -5.282572354
    2 │     2.52413358
    3 │    10.27538518
```

2 p-value

The p-value can be calculated from $p = 2 * (1 - T(\mathrm{df}, |t|))$, where T is the cumulative distribution function of Student's t distribution, df is the residual degrees of freedom, and $|t|$ is the absolute value of the observed t statistic. The t statistic was calculated before, and the residual degrees of freedom are returned as e(df_r). The absolute value can be calculated by using the abs() function, and $(1 - T(\mathrm{df}, t))$ can be calculated by using the ttail(df, t) function. The calculation is put together as follows:

```
. local t = _b[foreign]/_se[foreign]
. di 2*ttail(e(df_r),abs('t'))
.01383634
```

Using Mata, the vector of all p-values is then

```
: df = st_numscalar("e(df_r)")
: t = b :/ se
: 2*ttail(df, abs(t))
                    1

    1 │    1.33307e-06
    2 │    .0138363442
    3 │    1.08513e-15
```

3 Confidence interval

The lower and upper bounds of the confidence interval can be calculated as $\hat{b} \pm t_{\alpha/2}\mathrm{se}$, where $t_{\alpha/2}$ is the critical t-value given a significance level $\alpha/2$. This critical value can be calculated by using the invttail(df, $\alpha/2$) function. The lower and upper bounds of the 95% confidence interval for the parameter of foreign are thus given by

```
. di _b[foreign] - invttail(e(df_r),0.025)*_se[foreign]
371.2169
. di _b[foreign] + invttail(e(df_r),0.025)*_se[foreign]
3163.3676
```

The vectors of lower and upper bounds for all parameters follow suit in Mata as

```
: b :- invttail(df,0.025):*se, b :+ invttail(df,0.025):*se
                   1                2

    1 |  -405.2416661    -183.1494001
    2 |   371.2169028     3163.367584
    3 |     9595.1638     14215.66676
```

4   Models reporting z statistics

If you are using an estimation command that reports z statistics instead of t statistics, the values become

- _b[foreign]/_se[foreign] for the z statistic;

- 2*normal(-abs('z')) for the p-value (where the minus sign comes from the fact normal() starts with the lower tail of the distribution, whereas ttail() starts with the upper tail);

- _b[foreign] - invnormal(0.975)*_se[foreign] for the lower bound of the 95% confidence interval, and _b[foreign] + invnormal(0.975)*_se[foreign] for the upper bound (.975 is used instead of .025 for the same kind of reason).

5   Further comments

Often it is unnecessary to do these calculations. In particular, if you are interested in creating custom tables of regression-like output the estimates table command or the tools developed by Jann (2005, 2007) are much more convenient. Similarly, if the aim is to create graphs of regression output, take a good look at the tools developed by Newson (2003) before attempting to use the methods described here. This tip is for situations in which no such command does what you want.

References

Jann, B. 2005. Making regression tables from stored estimates. *Stata Journal* 5: 288–308.

———. 2007. Making regression tables simplified. *Stata Journal* 7: 227–244.

Newson, R. 2003. Confidence intervals and p-values for delivery to the end user. *Stata Journal* 3: 245–269.

The Stata Journal (2007)
7, Number 4, pp. 587–589

114

Stata tip 54: Post your results

Philippe Van Kerm
CEPS/INSTEAD
Differdange, Luxembourg
philippe.vankerm@ceps.lu

The command `post` and its companion commands `postfile` and `postclose` are described in [P] **postfile** as "utilities to assist Stata programmers in performing Monte Carlo type experiments". That description understates their usefulness, as `post` is one of the most flexible ways to accumulate results and save them for later use in an external file.

Stata output is displayed in the Results window and can be stored in log files. However, browsing log files and selecting particular results can be tedious and inefficient. Fortunately, there are several alternatives, including the use of `file` (see [P] **file**) or the `estimates` suite of commands (see [R] **estimates**), and `post`, the focus here.

Use of `post` is fully described in [P] **postfile**. The steps are in essence:

1. Call `postfile` to initialize the results file: identify the filename, name its variables, and determine their types.

2. Run the analysis and accumulate the results by repeatedly calling `post`. Each call to `post` adds one observation (record or line) to the results file.

3. Close the results file with `postclose`.

`post` is flexible in what it records: e-class, r-class, or s-class results, string or numeric values, locals, constants, etc. Posted results are recorded without disturbing the data in memory. This is particularly neat: it keeps datasets tidy and allows calling multiple files without interfering with the accumulation of results.

This first example uses the `auto` data. We loop over all possible combinations of `foreign` and `rep78` and save average `price` within each group. Estimates are recorded in a new file named `autoinfo.dta`, which is later opened for displaying results with `tabdisp`.

```
. tempname hdle
. postfile `hdle' foreign rep78 mean using autoinfo
. sysuse auto
(1978 Automobile Data)
. forvalues f=0/1 {
  2.          forvalues r=1/5 {
  3.                  summarize price if foreign==`f' & rep78==`r', meanonly
  4.                  post `hdle' (`f') (`r') (r(mean))
  5.          }
  6. }
. postclose `hdle'
```

```
. use autoinfo, clear
. label define lf 0 "Domestic car" 1 "Foreign car"
. label values foreign lf
. label variable foreign "Origin of car"
. label variable rep78 "1978 repair record"
. tabdisp rep78 foreign, cell(mean)
```

1978 repair record	Origin of car Domestic car	Foreign car
1	4564.5	
2	5967.625	
3	6607.074	4828.667
4	5881.556	6261.444
5	4204.5	6292.667

This example just shows the technique. In fact, for similar problems, the same effect can be produced easily with **statsby** (see [D] **statsby**):

```
. sysuse auto
(1978 Automobile Data)
. statsby mean=r(mean), by(foreign rep78) saving(autoinfo2): summarize price
(running summarize on estimation sample)
  (output omitted)
. use autoinfo2
(statsby: summarize)
. tabdisp rep78 foreign, cell(mean)
  (output omitted)
```

However, **statsby** is too restricted for more elaborate problems. A second example shows computations that store results for each of a series of files, here the numbers of observations and variables. It also demonstrates that graph commands are easily used for displaying results.

```
. tempname hdle
. postfile 'hdle' str20 name str100 label nobs nvar using sysfilesinfo
. sysuse dir
  (output omitted)
. local allfiles "'r(files)'"
. foreach dtafile of local allfiles {
  2.          sysuse 'dtafile', clear
  3.          describe, short
  4.          post 'hdle' ("'dtafile'") ('"'r: data label'"') (r(N)) (r(k))
  5. }
  (output omitted)
. postclose 'hdle'

. use sysfilesinfo
. keep if label!=""
(18 observations deleted)
```

```
. replace name = subinstr(name,".dta","",.)
(15 real changes made)
. label variable nobs "Number of observations in dataset"
. label variable nvar "Number of variables in dataset"
. scatter nvar nobs if nobs<250, mlabel(name) mlabpostion(12)
```

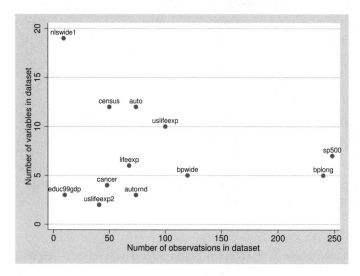

Figure 1: Exploiting posted results

The flexibility of `post` for both collection, when relevant results are posted, and processing, when collected results are analyzed, makes it useful in a broad range of settings, which is different from Monte Carlo simulations.

The Stata Journal (2007)
7, Number 4, pp. 590–592

Stata tip 55: Better axis labeling for time points and time intervals

Nicholas J. Cox
Department of Geography
Durham University
Durham City, UK
n.j.cox@durham.ac.uk

Plots of time-series data show time on one axis, usually the horizontal or x axis. Unless the number of time points is small, axis labels are usually given only for selected times. Users quickly find that Stata's default time axis labels are often not suitable for use in public. In fact, the most suitable labels may not correspond to *any* of the data points. This will arise when it is better to label longer time intervals, rather than any individual times in the dataset.

For example,

```
. webuse turksales
```

reads in 40 quarterly observations for 1990q1 to 1999q4 with a response variable of turkey sales. The default time axis labels with both `line sales t` and `tsline sales` are 1990q1, 1992q3, 1995q1, 1997q3, and 2000q1. These are not good choices for any purpose, even exploration of the data in private.

Label choice is partly a matter of taste, but you might well agree with Stata that labeling every time point would be busy and the result difficult to read. With 40 quarterly values, possible choices include one point per year (10 labels) and one point every other year (5 labels). One possibility is to label every fourth quarter, as that is usually the quarter with highest turkey sales. `summarize` reveals that the times range from 120 to 159 quarters (0 means the first quarter of 1960), so we can type

```
. line sales t, xlabel(123(4)159)
```

Note how we use a *numlist*, `123(4)159`, to avoid spelling out every value. The step length is 4 for four quarters. See [U] **11.1.8 numlist** or `help numlist` for more details of *numlist*s. This graph too would need more work before publication, as the labels are still crowded. The text of the labels (e.g., 1990q4) may or may not be judged suitable, depending partly on the readership for the graph.

However, there is another choice: label time intervals (years) and mark the boundaries between those time intervals by ticks. Consider 1990. The four quarters in Stata's units are 120, 121, 122, and 123. Thus we could put text showing the year at a midpoint of 121.5 and ticks showing year boundaries at 119.5 and 123.5. For all years, we should use the *numlist* idea again with the following command to produce figure 1.

```
. line sales t, xtick(119.5(4)159.5, tlength(*1.5))
> xlabel(121.5(4)157.5, noticks format(%tqCY)) xtitle("")
```

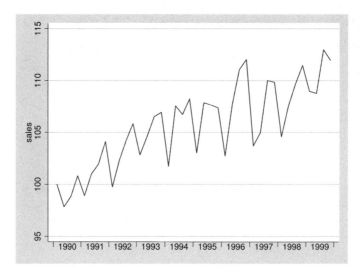

Figure 1: Turkey sales in each quarter. Time axis labels show years (with ticks suppressed) and time axis ticks show year ends.

The most important details here are suppressing the ticks for the axis labels and specifying a format for them. Cosmetic additions include lengthening the ticks compared with the default and suppressing the axis title, which would otherwise be the variable name `t` (or a variable label if it existed). It is usually clear from the labels what is being shown. Other possibilities include changing the text size for the axis label, changing the angle at which the axis label is shown, and suppressing the century by using a format like `%tqY`. Those may not be especially attractive, but nevertheless might be forced upon you by practicalities.

The main idea is clearly more general. The axis labels and the axis ticks need not correspond to each other, and it might be good to have fewer labels than ticks for longer series. Monthly and half-yearly data naturally yield to the same method, but use 12 or 2 and not 4 as the step length. Weekly and daily data are more awkward but still manageable.

If you were producing many similar graphs, you might want to automate this process to some degree. The mental arithmetic might easily be more challenging than in the turkey example. Let us imagine daily data for several years. Thus we could put ticks every January 1 and year labels every July 1. That will be adequate precision in practice. Find the first and last years in your data, if necessary by a command like `gen year = year(date)` followed by `summarize`. Suppose again that the years are 1990–1999. We can put the needed dates in local macros with a loop:

```
. forvalues y = 1990/1999 {
        local jan `jan' `=mdy(1,1,`y')'
        local jul `jul' `=mdy(7,1,`y')'
  }
```

Each time around the loop the daily dates for January 1 and July 1 in each year are calculated on the fly with a call to the `mdy()` function and added to a macro. For more details, see [P] **forvalues** and [P] **macro**, the corresponding help files, or Cox (2002). Once done, the graph command is something like

```
. line whatever date, xlabel('jul', format(%tdCY) noticks)
> xtick('jan', tlength(*1.5))
```

A key requirement is that the local macros used in the graph command must be visible, by virtue of being in the same interactive session, do-file, or program. That is in essence what `local` means.

Calendar years, meaning here Western calendar years, are clearly not the only possibilities. You could use other boundaries and midpoints for years or other periods defined by other criteria (e.g., academic, financial, fiscal, hydrological, political, religious).

Reference

Cox, N. J. 2002. Speaking Stata: How to face lists with fortitude. *Stata Journal* 2: 202–222.

The Stata Journal (2008)
8, Number 1, pp. 134–136

Stata tip 56: Writing parameterized text files

Rosa Gini
Regional Agency for Public Health of Tuscany
Florence, Italy
rosa.gini@arsanita.toscana.it

Stata includes several commands for text file manipulation. A good example is the copy command ([D] **copy**). Typing

```
. copy filename1 filename2
```

simply copies `filename1` to `filename2`, regardless of its content.

Often when dealing with text files, you need greater flexibility. Stata can also read and write text files using the `file` suite of commands ([P] **file**). Thereby, you can rewrite a text file by first reading and then writing it with modifications.

```
local filetarget "filename2"
local filesource "filename1"
local appendreplace "replace" /* append or replace */
tempname target source
file open `target' using `filetarget', write `appendreplace' text
file open `source' using `filesource', read text
file read `source' textline
while r(eof) == 0 {
        file write `target' `""`textline'""' _n
        file read `source' textline
}
file close `source'
file close `target'
```

A notable feature of this second way of copying text files is that you can append files to existing files. Even more importantly, you need not copy the file character for character. While rewriting, Stata may substitute the values of local or global macros that have been defined. This allows users to work with a template and produce text with elements substituted for each occasion. Such files may be called "parameterized", as they contain elements constant within a document but variable from document to document.

As an example of the many applications of this simple device, consider the needs of those who periodically access big databases to produce standard reports. The structure of the database is fixed; hence, access will be by a cascade of queries. The queries will be the same every time except for some parameters that change, such as the date (e.g., month, quarter, or year). Stata can access such databases without intermediaries, as SQL code for the queries can be stored in a text file with global macros and be rewritten and executed periodically using the odbc command ([D] **odbc**). A file `query_para.sql` might contain the following simple SQL parameterized code:

```
CREATE TABLE ${year}_AMI AS
SELECT H.ID, H.CODE_PATIENT, H.YEAR, H.SEX, H.AGE
FROM HOSPITALIZATIONS  H, PATHOLOGIES  H_PATHOL
WHERE H.ID=H_PATHOL.ID AND H_PATHOL.DIAGNOSIS="410" AND ${conditions} AND
> H.YEAR=${year} AND H_PATHOL.YEAR=${year};

CREATE INDEX ${year}_ID ON ${year}_AMI (CODE_PATIENT)
TABLESPACE epidemiology;
ANALYZE TABLE ${year}_AMI compute statistics;

CREATE TABLE ${year}_MORTALITY AS
SELECT DISTINCT CASES.CODE_PATIENT, MOR.DEATH_DATE
FROM ${year}_AMI CASES, MORTALITY  MOR
WHERE MOR.CODE_PATIENT=CASES.CODE_PATIENT;
```

This sequence of queries looks for patients hospitalized for AMI (acute myocardial infarction, or heart attack) in a given year and then links the list of patients to the mortality records to obtain data on survival. As the list of patients may be very long, the code computes an index to perform better linkage.

The following code is a template for one Stata session. For example, substitute any connect_options desired for odbc.

```
/* set parameters */
global year = 2005
global conditions "H.AGE>64"
/* rewrite text */
local filetarget "query.sql"
local filesource "query_para.sql"
local appendreplace "replace" /* append or replace */
tempname target source
file open `target' using `filetarget', write `appendreplace' text
file open `source' using `filesource', read text
local i = 1
file read `source' textline
while r(eof) == 0 {
        file write `target' `""`textline'""' _n
        local ++i
        file read `source' textline
}
file close `source'
file close `target'
/* execute queries */
odbc sqlfile("query.sql"), dsn("DataSourceName") [connect_options]
/* load and save generated tables */
foreach table in AMI MORTALITY {
        odbc load table("${year}_`table'"), clear [connect_options]
        save ${year}_`table', replace
}
```

The code will make Stata

1. Write a text file of actual (nonparameterized) SQL code, where ${year} is substituted by 2005 and ${conditions} is substituted by "H.AGE>64".

2. Execute the SQL code via odbc. This may take some time. If an SQL client is available, a practical alternative is to make Stata call that client and ask it

to execute the SQL using the `shell` command ([D] **shell**). This will make the execution of the queries independent of the Stata session.

3. Load the generated tables.

4. Save each of the generated tables as a Stata `.dta` file for later analysis.

The Stata Journal (2008)
8, Number 1, pp. 137–138

Stata tip 57: How to reinstall Stata

Bill Gould
StataCorp
College Station, TX
wgould@stata.com

Sometimes disaster, quite unbidden, may strike your use of Stata. Computers break, drives fail, viruses attack, you or your system administrator do silly—even stupid—things, and Stata stops working or, worse, only partially works, recognizing some commands but not others. What is the solution? You need to reinstall Stata. It will be easier than you may fear.

1. If Stata is still working, get a list of the user-written ado-files you have installed. Bring up your broken Stata and type

   ```
   . ado dir
   ```

   That should list the packages. `ado dir` is a built-in command of Stata, so even if the ado-files are missing, `ado dir` will work. Assuming it does work, let's type the following:

   ```
   . log using installed.log, replace
   . ado dir
   . log close
   ```

   Exit Stata and store the new file `installed.log` in a safe place. Print the file, too. In the worst case, we can use the information listed to reinstall the user-written files.

2. With both your original CD and your paper license handy, take a deep breath and reinstall Stata. If you cannot find your original license codes, call Stata Technical Services.

3. Launch the newly installed Stata. Type `ado dir`. That will either (1) list the files you previously had installed or (2) list nothing. Almost always, the result will be (1).

4. Regardless, `update` your Stata: Type `update query` and follow the instructions.

5. Now you are either done, or, very rarely, you still need to reinstall the user-written files. In that case, look at the original `ado dir` listing we obtained in step 1. One line might read

   ```
   [1]  package mf_invtokens from http://fmwww.bc.edu/RePEc/bocode/m
          'MF_INVTOKENS': module (Mata) to convert ...
   ```

   so you would type

   ```
   . net from http://fmwww.bc.edu/RePEc/bocode/m
   . net install mf_invtokens
   ```

In the case where the package is from http://fmwww.bc.edu, easier than the above is to type

```
. ssc install mf_invtokens
```

Both will do the same thing. `ssc` can be used only to install materials from http://fmwww.bc.edu/. In other cases, type the two `net` commands.

Anyway, work the list starting at the top.

Unless you have had a disk failure, it is exceedingly unlikely that you will lose the user-written programs. If you do not have a backup plan in place for your hard disk, it is a good idea to periodically log the output of `ado dir` and store the output in a safe place.

The Stata Journal (2008)
8, Number 1, pp. 139–141

Stata tip 58: nl is not just for nonlinear models

Brian P. Poi
StataCorp
College Station, TX
bpoi@stata.com

1 Introduction

The `nl` command makes performing nonlinear least-squares estimation almost as easy as performing linear regression. In this tip, three examples are given where `nl` is preferable to `regress`, even when the model is linear in the parameters.

2 Transforming independent variables

Using the venerable `auto` dataset, suppose we want to predict the weight of a car based on its fuel economy measured in miles per gallon. We first plot the data:

```
. sysuse auto
. scatter weight mpg
```

Clearly, there is a negative relationship between `weight` and `mpg`, but is that relationship linear? The engineer in each of us believes that the amount of gasoline used to go one mile should be a better predictor of weight than the number of miles a car can go on one gallon of gas, so we should focus on the reciprocal of `mpg`. One way to proceed would be to create a new variable, `gpm`, measuring gallons of gasoline per mile and then to use `regress` to fit a model of `weight` on `gpm`. However, consider using `nl` instead:

```
. nl (weight = {b0} + {b1}/mpg)
(obs = 74)
Iteration 0:  residual SS =  1.19e+07
Iteration 1:  residual SS =  1.19e+07
```

Source	SS	df	MS
Model	32190898.6	1	32190898.6
Residual	11903279.8	72	165323.33
Total	44094178.4	73	604029.841

	R-squared	=	0.7300
	Adj R-squared	=	0.7263
	Root MSE	=	406.5997
	Res. dev.	=	1097.134

weight	Coef.	Std. Err.	t	P>|t|	[95% Conf. Interval]	
/b0	415.1925	192.5243	2.16	0.034	31.40241	798.9826
/b1	51885.27	3718.301	13.95	0.000	44472.97	59297.56

Parameter b0 taken as constant term in model & ANOVA table

(You can verify that R^2 from this model is higher than that from a linear model of `weight` on `mpg`. You can also verify that our results match those from *regressing* `weight` on gpm.)

Here a key advantage of `nl` is that we do not need to create a new variable containing the reciprocal of `mpg`. When doing exploratory data analysis, we might want to consider using the natural log or square root of a variable as a regressor, and using `nl` saves us some typing in these cases. In general, instead of typing

```
. generate sqrtx = sqrt(x)
. regress y sqrtx
```

we can type

```
. nl (y = {b0} + {b1}*sqrt(x))
```

3   Marginal effects and elasticities[1]

Using `nl` has other advantages as well. In many applications, we include not just the variable x in our model but also x^2. For example, most wage equations express log wages as a function of experience and experience squared. Say we want to fit the model

$$y_i = \alpha + \beta_1 x_i + \beta_2 x_i^2 + \epsilon_i$$

and then determine the elasticity of y with respect to x; that is, we want to know the percent by which y will change if x changes by one percent.

Given the interest in an elasticity, the inclination might be to use the `mfx` command with the `eyex` option. We might type

```
. generate xsq = x^2
. regress y x xsq
. mfx compute, eyex
```

These commands will not give us the answer we expect because `regress` and `mfx` have no way of knowing that `xsq` is the square of `x`. Those commands just see two independent variables, and `mfx` will return two "elasticities", one for `x` and one for `xsq`. If x changes by some amount, then clearly x^2 will change as well; however, `mfx`, when computing the derivative of the regression function with respect to `x`, holds `xsq` fixed!

The easiest way to proceed is to use `nl` instead of `regress`:

```
. nl (y = {a} + {b1}*x + {b2}*x^2), variables(x)
. mfx compute, eyex
```

---

1. Editors' note: In Stata 11, this example can be done using `regress` with factor-variable notation and the new `margins` command:

```
. regress y x c.x#c.x
. margins, eyex(x) atmeans
```

However, this method only works with polynomials; more general nonlinear functions still require `nl`.

Whenever you intend to use `mfx` after `nl`, you must use the `variables()` option. This option causes `nl` to save those variable names among its estimation results.

4   Constraints

`nl` makes imposing nonlinear constraints easy. Say you have the linear regression model

$$y_i = \alpha + \beta_1 x_{1i} + \beta_2 x_{2i} + \beta_3 x_{3i} + \epsilon_i$$

and for whatever reason you want to impose the constraint that $\beta_2 \beta_3 = 5$. We cannot use the `constraint` command in conjunction with `regress` because `constraint` only works with linear constraints. `nl`, however, provides an easy way out. Our constraint implies that $\beta_3 = 5/\beta_2$, so we can type

```
. nl (y = {a} + {b1}*x1 + {b2=1}*x2 + (5/{b2})*x3)
```

Here we initialized β_2 to be 1 because if the product of β_2 and β_3 is not 0, then neither of those parameters can be 0, which is the default initial value used by `nl`.

The Stata Journal (2008)
8, Number 1, pp. 142–145

Stata tip 59: Plotting on any transformed scale

Nicholas J. Cox
Department of Geography
Durham University
Durham City, UK
n.j.cox@durham.ac.uk

Using a transformed scale on one or the other axis of a plot is a standard graphical technique throughout science. The most common example is the use of a logarithmic scale. This possibility is wired into Stata through options `yscale(log)` and `xscale(log)`; see [G] *axis_scale_options*. The only small difficulty is that Stata is not especially smart at reading your mind to discern what axis labels you want. When values range over several orders of magnitude, selected powers of 10 are likely to be convenient. When values range over a shorter interval, labels based on multiples of 1 2 5 10, 1 4 7 10, or 1 3 10 may all be good choices.

No other scale receives such special treatment in Stata. However, other transformations such as square roots (especially for counts) or reciprocals (e.g., in chemistry or biochemistry [Cornish-Bowden 2004]) are widely used in various kinds of plots. The aim of this tip is to show that plotting on *any* transformed scale is straightforward. As an example, we focus on logit scales for continuous proportions and percents.

Given proportions p, logit $p = \ln\{p/(1-p)\}$ is perhaps most familiar to many readers as a link function for binary response variables within logit modeling. Such logit modeling is now over 60 years old, but before that lies a century over which so-called logistic curves were used to model growth or decay in demography, ecology, physiology, chemistry, and other fields. Banks (1994), Kingsland (1995), and Cramer (2004) give historical details, many examples, and further references.

The growth of literacy and its complement—the decline of illiteracy—provide substantial examples. In a splendid monograph, Cipolla (1969) gives fascinating historical data but no graphs. Complete illiteracy and complete literacy provide asymptotes to any growth or decay curve, so even without any formal modeling we would broadly expect something like S-shaped or sigmoid curves. Logit scales in particular thus appear natural or at least convenient for plotting literacy data (Sopher 1974, 1979). More generally, plotting on logit scales goes back at least as far as Wilson (1925).

Figure 1 shows how many newly married people could not write their names in various countries during the late nineteenth century, as obtained with data from Cipolla (1969, 121–125) and the following commands:

```
. local yti "% newly married unable to write their names"
. line Italy_females Italy_males France_females France_males Scotland_females
> Scotland_males year, legend(pos(3) col(1) size(*0.8)) xla(1860(10)1900)
> xtitle("") yla(, ang(h)) ytitle(`yti')
```

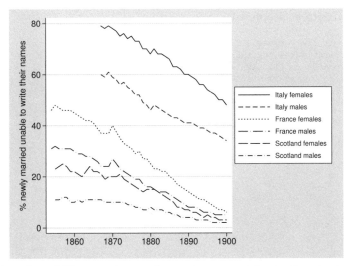

Figure 1: Line plot of illiteracy by sex for various countries in the nineteenth century.

How do we show such data on a logit scale? There are two steps. First, calculate the coordinates you want shown before you graph them. Here we loop over a bunch of variables and apply the transform logit(*percent*/100):

```
. foreach v of var *males {
.       gen logit_'v' = logit('v'/100)
.       label var logit_'v' "': var label 'v''"
. }
```

For tutorials on **foreach** and the machinery used here in looping, see Cox (2002, 2003). Note that, at the same time, we copy variable labels across so that they will show up automatically on later graph legends.

Second, and just slightly more difficult, is to get axis labels as we want them (and axis ticks also, if needed). Even people who work with logits all the time usually do not want to decode that a logit of 0 means 50%, or a logit of 1 means 73.1%, and so forth, even if the **invlogit()** function makes the calculation easy. Logit scales stretch percents near 0 or 100 compared with those near 50. Inspection of figure 1 suggests that 2 5 10(10)80 would be good labels to show for percents within the range of the data. So we want text like 50 to be shown where the graph is showing logit(50/100). The key trick is to pack all the text we want to show and where that text should go into a local macro.

```
. foreach n of num 2 5 10(10)80 {
.       local label 'label' '= logit('n'/100)' "'n'"
. }
```

To see what is happening, follow the loop: First time around, local macro 'n' takes on the value 2. logit(2/100) is evaluated on the fly (the result is about −3.8918) and that is where on our *y* axis the text "2" should go. Second time around, the same is done for 5 and logit(5/100). And so forth over the numlist 2 5 10(10)80.

Now we can get our graph with logit scale:

```
. line logit_Italy_females logit_Italy_males logit_France_females
> logit_France_males logit_Scotland_females logit_Scotland_males year,
> legend(pos(3) col(1) size(*0.8)) xla(1860(10)1900) xtitle("")
> yla(`label', ang(h)) ytitle(`yti')
```

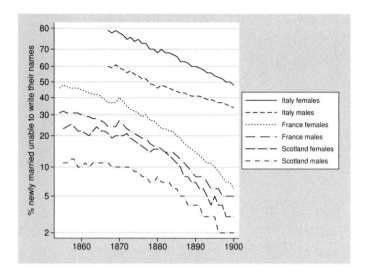

Figure 2: Line plot of illiteracy by sex for various countries in the nineteenth century. Note the logit scale for the response.

Specifically in this example, we now can see data on a more natural scale, complementing the original raw scale. The granularity of the data (rounded to integer percents) is also evident.

Generally, some small details of macro handling deserve flagging. You may be accustomed to a tidy form of local macro definition:

```
. local macname "contents"
```

But the delimiters " " used here would, for this problem, complicate processing given that we do want double quotes inside the macro. Note from the previous `foreach` loop that they can be left off, to advantage.

In practice, you might need to iterate over several possible sets of labels before you get the graph you most like. Repeating the whole of the `foreach` loop would mean that the local macro would continue to accumulate material. Blanking the macro out with

```
. local label
```

will let you start from scratch.

The recipe for ticks is even easier. Suppose we want ticks at 15(10)75, that is, at 15 25 35 45 55 65 75. We just need to be able to tell Stata exactly where to put them:

```
. foreach n of num 15(10)75 {
.       local ticks `ticks' `= logit(`n'/100)'
. }
```

Then specify an option such as `yticks(`ticks')` in the graph command.

Finally, note that the local macros you define must be visible to the graph command you issue, namely within the same interactive session, do-file, or program. That is what local means, after all.

In a nutshell: Showing data on any transformed scale is a matter of doing the transformation in advance, after which you need only to fix axis labels or ticks. The latter is best achieved by building graph option arguments in a local macro.

References

Banks, R. B. 1994. *Growth and Diffusion Phenomena: Mathematical Frameworks and Applications.* Berlin: Springer.

Cipolla, C. M. 1969. *Literacy and Development in the West.* Harmondsworth: Penguin.

Cornish-Bowden, A. 2004. *Fundamentals of Enzyme Kinetics.* 3rd ed. London: Portland Press.

Cox, N. J. 2002. Speaking Stata: How to face lists with fortitude. *Stata Journal* 2: 202–222.

———. 2003. Speaking Stata: Problems with lists. *Stata Journal* 3: 185–202.

Cramer, J. S. 2004. The early origins of the logit model. *Studies in History and Philosophy of Biological and Biomedical Sciences* 35: 613–626.

Kingsland, S. E. 1995. *Modeling Nature: Episodes in the History of Population Ecology.* 2nd ed. Chicago: University of Chicago Press.

Sopher, D. E. 1974. A measure of disparity. *Professional Geographer* 26: 389–392.

———. 1979. Temporal disparity as a measure of change. *Professional Geographer* 31: 377–381.

Wilson, E. B. 1925. The logistic or autocatalytic grid. *Proceedings of the National Academy of Sciences* 11: 451–456.

Stata tip 60: Making fast and easy changes to files with filefilter

Alan R. Riley
StataCorp
College Station, TX
ariley@stata.com

Stata has a command `filefilter` (see [D] **filefilter**) that makes it possible to perform global search-and-replace operations on a file, saving the result to another file. Think of it as a command which copies a file, but in the process of copying it, can search for one text pattern and replace it with another.

As its manual entry points out, `filefilter` has been designed to read and write the input and output files using buffers for speed, and is thus fast at converting even large files. `filefilter` can be used on files which are too large to open in a traditional text editor, and because Stata is programmable, it is possible to use `filefilter` to perform complicated global search-and-replace operations, which would not be possible in most text editors. `filefilter` can even make changes to binary files.

`filefilter` is often used to preprocess files (perhaps to remove invalid characters or to change a delimiter) before reading them into Stata and can be used in many other situations as a useful file-processing tool.

For example, if you have a log file named `x.log` in Windows (where the end-of-line (EOL) character combination is \r\n), and you want to convert the file to have Unix-style EOL characters (\n), you can type in Stata

```
. filefilter x.log y.log, from(\r\n) to(\n)
```

which will replace every occurrence of the Windows EOL character sequence with the Unix EOL character. Equivalently, you could type

```
. filefilter x.log y.log, from(\W) to(\U)
```

because `filefilter` understands \W as a synonym for the Windows EOL character sequence \r\n and \U as a synonym for the Unix EOL character sequence \n.

(For the rest of this tip, I will write \W as the EOL marker, but be sure to use the EOL shorthand in `filefilter` appropriate for your operating system.)

Let's put `filefilter` to use on another example. Imagine that we want to replace all occurrences of multiple blank lines in a file with a single blank line for readability. Changing a file in this way may be desirable after, for example, the command `cleanlog` (Sieswerda 2003) which reads a log file (plain text or SMCL) and removes all command syntax and other extraneous material, leaving behind only output. However, in doing so, `cleanlog` leaves behind multiple adjacent blank lines.

Consider the following lines from a file. (The file below obviously did not result from `cleanlog`, but it will serve for the purpose of this example.) I will write EOL everywhere the file contains an end-of-line character sequence.

```
here is a line.  the next two lines are blank in the original file.EOL
EOL
EOL
here is another line.  the next line is blank in the original file.EOL
EOL
this is the last line of the file.EOL
```

The first `filefilter` syntax you might think of would be

```
. filefilter x.log y.log, from(\r\n\r\n) to(\r\n)
```

but it will not do what we want. Because there are EOL characters at the end of nonblank lines, if all adjacent pairs of EOL characters (`\W\W`) were replaced with single EOL characters (`\W`), the file above would end up looking like

```
here is a line.  the next two lines are blank in the original file.EOL
here is another line.  the next line is blank in the original file.EOL
this is the last line of the file.EOL
```

with no blank lines at all. To have a blank line between sections of output, there must be two adjacent EOL characters: one at the end of a line, and another on a line all by itself (the blank line).

Thus, to compress multiple adjacent blank lines down to single blank lines, we need to replace every occurrence of three adjacent EOL characters with two EOL characters:

```
. filefilter x.log y.log, from(\W\W\W) to(\W\W)
```

We still have not quite achieved the desired result. If we issue the above command only once, there may still be adjacent empty lines left in the file. We actually need to call `filefilter` multiple times, each time changing every three newlines to two newlines. I will assume that `x.log` is the original file, and will use `y.log` and `z.log` as output files with `filefilter` so that the original file will be left unchanged:

```
. filefilter x.log y.log, from(\W\W\W) to(\W\W)
. filefilter y.log z.log, from(\W\W\W) to(\W\W)
. filefilter z.log y.log, from(\W\W\W) to(\W\W) replace
. filefilter y.log z.log, from(\W\W\W) to(\W\W) replace
. filefilter z.log y.log, from(\W\W\W) to(\W\W) replace
...
```

The above should continue until no more changes are made. We can automate this by checking the return results from `filefilter` to see if the `from()` pattern was found. If it was not, we know there were no changes made, and thus, no more changes to be made:

```
filefilter x.log y.log, from(\W\W\W) to(\W\W)
local nchanges = r(occurrences)
while `nchanges' != 0 {
    filefilter y.log z.log, from(\W\W\W) to(\W\W) replace
    filefilter z.log y.log, from(\W\W\W) to(\W\W) replace
    local nchanges = r(occurrences)
}
...
```

After the code above is executed, `y.log` will contain the desired file, and `z.log` can be discarded. It is possible that the code above will call `filefilter` one more time than is necessary, but unless we have an extremely large file that takes `filefilter` some time to process, we won't even notice.

While it may seem inefficient to use `filefilter` to make multiple passes through a file until the desired result is achieved, it is a fast and easy way to make such modifications. For very large files, Stata's `file` command ([P] **file**) or Mata's I/O functions ([M-4] **io**) could be used to perform such processing in a single pass, but they require a higher level of programming effort.

Reference

Sieswerda, L. E. 2003. cleanlog: Stata module to clean log files. Boston College Department of Economics, Statistical Software Components S432401. Downloadable from http://ideas.repec.org/c/boc/bocode/s432401.html.

The Stata Journal (2008)
8, Number 2, pp. 293–294

Stata tip 61: Decimal commas in results output and data input

Nicholas J. Cox
Department of Geography
Durham University
Durham City, UK
n.j.cox@durham.ac.uk

Given a decimal fraction to evaluate, such as 5/4, Stata by default uses a period (stop) as a decimal separator and shows the result as 1.25. That is, the period separates, and also joins, the integer part 1 and the fractional part 25, meaning here 25/100. Many Stata users, particularly in the United States and several other English-speaking countries, will have learned of such decimal points at an early age and so may think little of this. However, in many other countries, commas are used as decimal separators, so that 1,25 is the preferred way of representing such fractions. This tip is for those users, although it also provides an example of how different user preferences can be accommodated by Stata.

Over several centuries, mathematicians and others have been using decimal fractions without uniformity in their representation. Cajori (1928, 314–335) gave one detailed historical discussion that is nevertheless incomplete. Periods and commas have been the most commonly used separators since the development of printing. Some authors, perhaps most notably John Napier of logarithm fame, even used both symbols in their work. An objection to the period is its common use to indicate multiplication, so some people have preferred centered or even raised dots, even though the same objection can be made in reverse, at least to centered dots. An objection to the comma is similarly to its common use to separate numbers in lists, although serious ambiguity should not arise so long as such lists are suitably spaced out.

Incidentally, many notations other than periods and commas have been proposed and several remained in use until well into the twentieth century. Indeed the momayyez, a mark like a forward slash or comma, is widely used at present in several Middle Eastern countries.

Stata 7 introduced `set dp comma` as a way to set the decimal point to a comma in output. Thus after

```
. set dp comma
. display 5/4
```

shows 1,25 and other output follows the same rule. Type `set dp comma, permanently` to have such output permanently. Type `set dp period` to restore the default.

Stata 7 also introduced comma-based formats such as `%7,2f`. See the help or manual entry for `format` for more details.

This still leaves the large question of input. Suppose, for example, that you have text or other data files in which numeric variables are indicated with commas as separators. Stata will not accept such variables as numeric, but the solution now is simple. Read in such variables as string, and then within Stata use `destring, replace dpcomma`. `destring` is especially convenient because it can be applied to several variables at once.

The `dpcomma` option was added to the `destring` command on 15 October 2007 and so is not documented in the Stata 10 manuals and not implemented in versions earlier than 10. Users still using Stata 9 or earlier are advised to use the `subinstr()` function to change commas to periods, followed by `destring`. Naturally, great care is required if any periods also appear as separators before or after the decimal comma. Any such periods should be removed before the comma is converted to a period.

Reference

Cajori, F. 1928. *A History of Mathematical Notations. Volume I: Notation in Elementary Mathematics*. Chicago: Open Court.

The Stata Journal (2008)
8, Number 2, pp. 295–298

Stata tip 62: Plotting on reversed scales

Nicholas J. Cox and Natasha L. M. Barlow
Durham University
Durham City, UK
n.j.cox@durham.ac.uk and n.l.m.barlow@durham.ac.uk

Stata has long had options allowing a reversed scale on either the y or the x axis of many of its graph types. Many graph users perhaps never even consider specifying such options. Those who do need to reach for them may wish to see detailed examples of how reversed scales may be exploited to good effect.

The usual Cartesian conventions are that vertical or y scales increase upward, from bottom to top, and horizontal or x scales increase from left to right. Vertical scales increasing downward are needed for graphs with vertical categorical axes following a table-like convention, in which the first (lowest) category is at the top of a graph, just as it would be at the top of a table. Commands such as `graph bar` and `graph dot` follow this convention. Indeed, it is likely to be so familiar that you may have to reflect briefly to see that is how such graphs are drawn. Other examples of this principle are discussed elsewhere in this issue (Cox 2008).

Reversed scales are also common in the Earth and environmental sciences. Here, in fields such as pedology, sedimentology, geomorphology, limnology, and oceanography, it is common to take measurements at varying depths within soils, sediments, rocks, and water bodies. Even though depth is not a response variable, it is conventional and convenient to plot depth below surface on the vertical axis; hence, the need for a reversed scale. An extra twist that gives spin to the graph problem is that frequently detailed analyses of materials at each level in a core, bore, or vertical profile yield several response variables, which are all to be plotted on the horizontal axis. Such multiple plotting is easiest when overlay is possible.

Let us look at some specific syntax for an example and then add comments. The data shown here come from work in progress by the second author. Sediment samples at 2 cm intervals down a core from Girdwood, Alaska, were examined for several elements associated with placer mining pollution (LaPerriere, Wagener, and Bjerklie 1985). In this example, the concentrations of gold, cadmium, arsenic, lead, copper, and zinc, measured in parts per million (ppm), are then plotted as a function of depth (cm). See figure 1.

(Continued on next page)

```
. local spec clwidth(medium) msize(*0.8)

. twoway
> connect depth Au, 'spec' ms(Oh) cmissing(n) ||
> connect depth Cd, 'spec' ms(Th)   ||
> connect depth As, 'spec' ms(Sh)   ||
> connect depth Pb, 'spec' ms(O)    ||
> connect depth Cu, 'spec' ms(T)    ||
> connect depth Zn, 'spec' ms(S)
> yscale(reverse)  xscale(log) xscale(titlegap(*10))
> ylabel(50(10)90, angle(h)) xlabel(0.1 0.3 1 3 10 30 100)
> ytitle(Depth (cm)) xtitle(Concentration (ppm))
> legend(order(1 "Au" 2 "Cd" 3 "As" 4 "Pb" 5 "Cu" 6 "Zn") position(3) column(1))
```

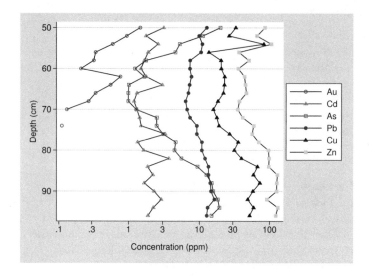

Figure 1: Variation in concentrations of various elements with depth, measured at a site in Girdwood, Alaska.

A key detail here is that Stata's model for scatter and similar plots is asymmetric. One or more y variables are allowed, but only one x variable, in each individual plot. Thus, if you wish to have several x variables, you must either superimpose several variables, as here, or juxtapose several plots horizontally.

Further, it is necessary to spell out what is desired as text in the legend. By default, twoway would use information for the y axis variables, which is not what is wanted for this kind of graph.

For these data a logarithmic scale is helpful, indeed essential, for showing several elements that vary greatly in abundance. Gold is reported as less than 0.1 ppm at depth. Such censored values cannot be shown by point symbols.

Even very experienced Stata users will not usually think up an entire graph command like this at the outset. Typically, you start with a fairly simple design and then elaborate it by a series of very small changes (which may well be changes of mind back and forth).

At some point, you are likely to find yourself transferring from the Command window to the Do-file Editor and from an interactive session to a do-file. We also find it helpful, once out of the Command window, to space out a command so that its elements are easier to see and so that it is easier to edit. Spending a few moments doing that saves some fiddly work later on. What is shown above is in fact more compressed than what typically appears within the Do-file Editor in our sessions.

The legend used here is, like all legends, at best a necessary evil, as it obliges careful readers to scan back and forth repeatedly to see what is what. The several vertical traces are fairly distinct, so one possibility is to extend the vertical axis and insert legend text at the top of the graph. '=Au[1]', for example, instructs Stata to evaluate the first value of Au and use its value. The same effect would be achieved by typing in the actual value. Here we are exploiting the shortness of the element names, which remain informative to any reader with minimal chemical knowledge. The legend itself can then be suppressed. The text elements could also be repeated at the bottom of the graph if desired. See figure 2.

```
. twoway
> connect depth Au, 'spec' ms(Oh) cmissing(n) ||
> connect depth Cd, 'spec' ms(Th)   ||
> connect depth As, 'spec' ms(Sh)   ||
> connect depth Pb, 'spec' ms(O)    ||
> connect depth Cu, 'spec' ms(T)    ||
> connect depth Zn, 'spec' ms(S)
> yscale(reverse r(46 .)) xscale(log) xscale(titlegap(*10))
> ylabel(50(10)90, angle(h)) xlabel(0.1 0.3 1 3 10 30 100)
> ytitle(Depth (cm)) xtitle(Concentration (ppm))
> text(48 '=Au[1]' "Au" 48 '=Cd[1]' "Cd" 48 '=As[1]' "As" 48 '=Pb[1]' "Pb"
>      48 '=Cu[1]' "Cu" 48 '=Zn[1]' "Zn")
> legend(off)
```

(Continued on next page)

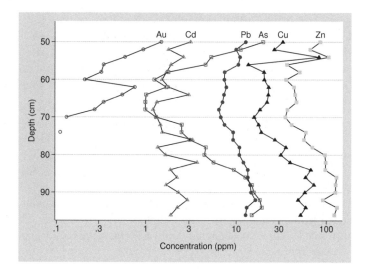

Figure 2: Variation in concentrations of various elements with depth, measured at a site in Girdwood, Alaska. The legend has been suppressed and its text elements placed within the graph.

The structure of overlays may suggest that a program be written that takes one y and several x variables and then works out the overlay code for you. Such a program might then be available particularly to colleagues and students less familiar with Stata. That idea is more tempting in this example than in general. In other cases, there might be a desire to mix different `twoway` types, to use different marker colors, or to make any number of other changes to distinguish the different variables. Any code flexible enough to specify any or all of that would lead to commands no easier to write than what is already possible. In computing as in cookery, sometimes you just keep old recipes once you have worked them out, at least as starting points for some later problem.

In the Earth and environmental sciences, reversed horizontal scales are also common for showing time, whenever the units are not calendar time, measured forward, but time before the present, measured backward. Within Stata, such practice typically requires no more than specifying `xscale(reverse)`.

References

Cox, N. J. 2008. Speaking Stata: Between tables and graphs. *Stata Journal* 8: 269–289.

LaPerriere, J. D., S. M. Wagener, and D. M. Bjerklie. 1985. Gold-mining effects on heavy metals in streams, Circle Quadrangle, Alaska. *Journal of the American Water Resources Association* 21: 245–252.

The Stata Journal (2008)
8, Number 2, pp. 299–303

Stata tip 63: Modeling proportions

Christopher F. Baum
Department of Economics
Boston College
Chestnut Hill, MA
baum@bc.edu

You may often want to model a response variable that appears as a proportion or fraction: the share of consumers' spending on food, the fraction of the vote for a candidate, or the fraction of days when air pollution is above acceptable levels in a city. To handle these data properly, you must take account of the bounded nature of the response. Just as a linear probability model on unit record data can generate predictions outside the unit interval, using a proportion in a linear regression model will generally yield nonsensical predictions for extreme values of the regressors.

One way to handle this for response variables' values strictly within the unit interval is the logit transformation

$$y = \frac{1}{1 + \exp(-X\beta)}$$

which yields the transformed response variable y^*

$$y^* = \log\left(\frac{y}{1-y}\right) = X\beta + \epsilon$$

where we have added a stochastic error process ϵ to the model to be fitted. This transformation may be performed with Stata's `logit()` function. We can then use linear regression ([R] **regress**) to model y^*, the *logit transformation* of y, as a linear function of a set of regressors, X. If we then generate predictions for our model ([R] **predict**), we can apply Stata's `invlogit()` function to express the predictions in units of y. For instance,

```
. use http://www.stata-press.com/data/r10/census7
(1980 Census data by state)
. generate adultpop = pop18p/pop
. quietly tabulate region, generate(R)
. generate marrate = marriage/pop
. generate divrate = divorce/pop
. generate ladultpop = logit(adultpop)
```

```
. regress ladultpop marrate divrate R1-R3
```

Source	SS	df	MS
Model	.164672377	5	.032934475
Residual	.327373732	43	.007613343
Total	.492046109	48	.010250961

Number of obs =	49
F(5, 43) =	4.33
Prob > F =	0.0028
R-squared =	0.3347
Adj R-squared =	0.2573
Root MSE =	.08725

ladultpop	Coef.	Std. Err.	t	P>|t|	[95% Conf.	Interval]
marrate	-18.26494	7.754941	-2.36	0.023	-33.90427	-2.625615
divrate	8.600844	12.09833	0.71	0.481	-15.79777	32.99946
R1	.1192464	.0482428	2.47	0.017	.0219555	.2165373
R2	.0498657	.042209	1.18	0.244	-.0352569	.1349883
R3	.0582061	.0357729	1.63	0.111	-.0139368	.130349
_cons	.999169	.093568	10.68	0.000	.8104712	1.187867

```
. predict double ladultpophat, xb

. generate adultpophat = invlogit(ladultpophat)

. summarize adultpop adultpophat
```

Variable	Obs	Mean	Std. Dev.	Min	Max
adultpop	49	.7113268	.0211434	.6303276	.7578948
adultpophat	49	.7116068	.0120068	.6855482	.7335103

Alternatively, we could use Stata's grouped logistic regression ([R] **glogit**) to fit the model. This command uses the same transformation on the response variable, which must be provided for the number of positive responses and the total number of responses (that is, the numerator and denominator of the proportion). For example,

```
. glogit pop18p pop marrate divrate R1-R3
Weighted LS logistic regression for grouped data
```

Source	SS	df	MS
Model	.129077492	5	.025815498
Residual	.261736024	43	.006086884
Total	.390813516	48	.008141948

Number of obs =	49
F(5, 43) =	4.24
Prob > F =	0.0032
R-squared =	0.3303
Adj R-squared =	0.2524
Root MSE =	.07802

pop18p	Coef.	Std. Err.	t	P>|t|	[95% Conf.	Interval]
marrate	-22.82454	8.476256	-2.69	0.010	-39.91853	-5.730537
divrate	18.44877	12.66291	1.46	0.152	-7.088418	43.98596
R1	.0762246	.0458899	1.66	0.104	-.0163212	.1687704
R2	-.0207864	.0362001	-0.57	0.569	-.0937909	.0522181
R3	.0088961	.0354021	0.25	0.803	-.062499	.0802912
_cons	1.058316	.0893998	11.84	0.000	.8780241	1.238608

These results differ from those of standard regression because `glogit` uses weighted least-squares techniques. As explained in [R] **glogit**, the appropriate weights correct for the heteroskedastic nature of ϵ has zero mean but variance equal to

$$\sigma_j^2 = \frac{1}{n_j p_j (1 - p_j)}$$

By generating those weights, where n_j is the number of responses in the jth category and p_j is the predicted value we computed above, we can reproduce the `glogit` results with `regress` by using analytic weights, as verified with the commands:

```
. generate glswt = adultpophat * (1 - adultpophat) * pop
. quietly regress ladultpop marrate divrate R1-R3 [aw=glswt]
```

In the case of these state-level census data, values for the proportion y must lie within the unit interval. But we often consider data for which the limiting values of zero or one are possible. A city may spend 0% of its budget on preschool enrichment programs. A county might have zero miles of active railway within its boundaries. There might have been zero murders in a particular town in each of the last five years. A hospital may have performed zero heart transplants last year. In other cases, we may find values of one for particular proportions of interest. Neither zeros nor ones can be included in the strategy above, as the logit transformation is not defined for those values.

A strategy for handling proportions data in which zeros and ones may appear as well as intermediate values was proposed by Papke and Wooldridge (1996). At the time of their writing, Stata's generalized linear model ([R] **glm**) command could not handle this model, but it has been enhanced to do so. This approach makes use of the logit link function (that is, the logit transformation of the response variable) and the binomial distribution, which may be a good choice of family even if the response is continuous. The variance of the binomial distribution must go to zero as the mean goes to either 0 or 1, as in each case the variable is approaching a constant, and the variance will be maximized for a variable with mean of 0.5.

To illustrate, consider an alternative dataset that contains zeros and ones in its response variable, `meals`: the proportion of students receiving free or subsidized meals at school.

(Continued on next page)

```
. use http://www.ats.ucla.edu/stat/stata/faq/proportion, clear
. summarize meals
```

Variable	Obs	Mean	Std. Dev.	Min	Max
meals	4421	.5188102	.3107313	0	1

```
. glm meals yr_rnd parented api99, link(logit) family(binomial) vce(robust) nolog
note: meals has noninteger values
```

Generalized linear models			No. of obs	=	4257
Optimization	: ML		Residual df	=	4253
			Scale parameter =		1
Deviance	= 395.8141242		(1/df) Deviance =		.093067
Pearson	= 374.7025759		(1/df) Pearson =		.0881031
Variance function: V(u) = u*(1-u/1)			[Binomial]		
Link function	: g(u) = ln(u/(1-u))		[Logit]		
			AIC	=	.7220973
Log pseudolikelihood = -1532.984106			BIC	=	-35143.61

meals	Coef.	Robust Std. Err.	z	P>|z|	[95% Conf. Interval]	
yr_rnd	.0482527	.0321714	1.50	0.134	-.0148021	.1113074
parented	-.7662598	.0390715	-19.61	0.000	-.8428386	-.6896811
api99	-.0073046	.0002156	-33.89	0.000	-.0077271	-.0068821
_cons	6.75343	.0896767	75.31	0.000	6.577667	6.929193

The techniques used above can be used to generate predictions from the model and transform them back into the units of the response variable. This approach is preferred to that of dropping the observations with zero or unit values, which would create a truncation problem, or coding them with some arbitrary value ("winsorizing") such as 0.0001 or 0.9999.

Some researchers have considered using censored normal regression techniques such as tobit ([R] **tobit**) on proportions data that contain zeros or ones. However, this is not an appropriate strategy, as the observed data in this case are not censored: values outside the $[0, 1]$ interval are not feasible for proportions data.

One concern was voiced about proportions data containing zeros or ones.[1] In the context of the generalized tobit or "heckit" model ([R] **heckman**), we allow for limit observations (for instance, zero values) being generated by a different process than non-censored observations. The same argument may apply here, in the case of proportions data: the managers of a city that spends none of its resources on preschool enrichment programs have made a discrete choice. A hospital with zero heart transplants may be a facility whose managers have chosen not to offer certain advanced services.

In this context, the glm approach, while properly handling both zeros and ones, does not allow for an alternative model of behavior generating the limit values. If different factors generate the observations at the limit points, a sample selection issue arises. Li and Nagpurnanand (2007) argue that selection issues arise in numerous variables of interest in corporate finance research. In a forthcoming article, Cook, Kieschnick, and

---

1. See, for instance, McDowell and Cox (2001).

McCullough (2008) address this issue for proportions of financial variables by developing what they term the "zero-inflated beta" model, which allows for zero values (but not unit values) in the proportion and for separate variables influencing the zero and nonzero values.[2]

References

Cook, D. O., R. Kieschnick, and B. D. McCullough. 2008. Regression analysis of proportions in finance with self selection. *Journal of Empirical Finance.* In press.

Li, K., and R. Nagpurnanand. 2007. Self-selection models in corporate finance. In *Handbook of Corporate Finance: Empirical Corporate Finance*, ed. B. E. Eckbo, chap. 2. Amsterdam: Elsevier.

McDowell, A., and N. J. Cox. 2001. FAQ: How do you fit a model when the dependent variable is a proportion? http://www.stata.com/support/faqs/stat/logit.html.

Papke, L. E., and J. M. Wooldridge. 1996. Econometric methods for fractional response variables with an application to 401(K) plan participation rates. *Journal of Applied Econometrics* 11: 619–632.

---

2. Their approach generalizes models fit with the beta distribution; user-written programs for that purpose may be located by typing `findit beta distribution`.

The Stata Journal (2008)
8, Number 3, pp. 444–445

Stata tip 64: Cleaning up user-entered string variables

Jeph Herrin
Yale School of Medicine
Yale University
New Haven, CT
jeph.herrin@yale.edu

Eva Poen
School of Economics
University of Nottingham
Nottingham, UK
eva.poen@gmail.com

A common problem in data management, especially when using large databases that receive entries from many different people, is that the same name is given in several different forms. This problem can arise in many instances, for example, lists of names of schools, hospitals, drugs, companies, countries, and so forth. Variation can reflect several genuinely different forms of the same name as well as a multitude of small errors or idiosyncrasies in spelling, punctuation, spacing, and use of uppercase or lowercase.

Thus, in context, a person may have no difficulty in recognizing that values of a string variable, such as "New York", "New York City", "N Y C", and so on, all mean the same thing. However, a program like Stata is necessarily literal and will treat them as distinct. How do we massage data so that values with the same meaning are represented in the same way? Several techniques exist for these purposes. Here we outline a simple strategy for ensuring that names are as tidy as possible.

As a preliminary stage, it is useful to try to eliminate small inconsistencies before you look at the individual observations. A good tactic is to keep the original names in one variable, exactly as given, and to work with one or more variables that contain cleaned-up versions. Some common problems are the following:

- Leading and trailing spaces may not be evident but will cause Stata to treat values as distinct. Thus "New York City" and "New York City " are not considered equal by Stata until `trim()` is used to delete the trailing space.

- Similarly, inconsistencies in internal spacing can cause differences that Stata will register. The `itrim()` function will reduce multiple, consecutive internal blanks to single internal blanks.

- Variations of uppercase and lowercase can also be troublesome. The `upper()`, `lower()`, or `proper()` functions can be used to make names consistent.

- Other common differences include whether hyphens are present, whether accented characters appear with or without accents or in some other form, and whether ampersands are printed as characters or as equivalent words.

- A large class of problems concerns abbreviations.

In the last two cases, `subinstr()` is a useful function for making changes toward consistent conventions. Note a common element here: string functions, documented in [D] **functions**, are invaluable for cleaning up strings.

After a preliminary cleaning, you can create a list of all the names that you have. Usually, this list is shorter than the number of observations. It is worthwhile to inspect the list of names and look for further clean-up possibilities before proceeding. A tabulation of names, say, by using `tabulate, sort`, serves this purpose and provides you with the number of occurrences for each variation.

After the cleaning is completed, you are ready to compile your list of names. Suppose that the `name` variable contains the names.

```
. use mydatafile
. by name, sort: keep if _n == 1
. keep name
```

Now open the Data Editor by typing

```
. edit name
```

and add a second—numeric—variable, say, `code`. In this second variable, give the same number for every observation that represents the same object. This will be moderately time consuming, but because data are sorted on `name`, it may just take a few minutes.

Now exit the Data Editor, sort on `name`, and save the dataset:

```
. sort name
. save codes, replace
```

Some people may prefer to create the code in their favorite spreadsheet or text editor, say, if an outside expert not adept at Stata is recruited to do the coding. The principles are the same: you need to export the data to the other application and then read data back into Stata. You may lose out on an audit trail if the other software does not offer an equivalent to a Stata `.log` file.

Now you have a file (`codes.dta`) that has a list of the names in all their variety, as well as a set of numeric codes, which you can return to later to check your work. The key thing is that you can now `merge` this file into your original file to assign a common code to every value of `name` that is the same:

```
. use mydatafile, clear
. sort name
. merge name using codes
```

As always when using `merge`, examine the `_merge` variable; here, if `_merge` is not always equal to 3, then you have made a mistake somewhere. You should also examine `code`; if there are any missing values, you will need to `edit` the file `codes.dta` again to add them.

Now you can identify the objects by their codes; if you want, you can assign a common name:

```
. by code, sort: replace name = name[1]
```

The Stata Journal (2008)
8, Number 3, pp. 446–447

Stata tip 65: Beware the backstabbing backslash

Nicholas J. Cox
Department of Geography
Durham University
Durham City, UK
n.j.cox@durham.ac.uk

The backslash character, \, has two main roles for Stata users who have Microsoft Windows as their operating system. This tip warns you to keep these roles distinct, because Stata's interpretation of what you want may puzzle you. The problem is signaled at [U] **18.3.11 Constructing Windows filenames using macros** but nevertheless bites often enough that another warning may be helpful.

The first and better known role is that the backslash acts as a separator in full specifications of directory or filenames. Thus, on many Windows machines, a Stata executable may be found within `C:\Program Files\Stata10`. Note the two backslashes in this example.

The second and lesser known role is that the backslash is a so-called escape character that suppresses the default interpretation of a character.

The best example in Stata is that the left quotation mark, `, is used to delimit the start of local macro names. Thus `` `frog' `` to Stata is a reference to a local macro called `frog`. Typically, Stata sees such a reference and substitutes the contents of the local macro `frog` at the same place. If no such macro is visible, that is not a bug. Instead, Stata substitutes an empty string for the macro name.

What happens if you want the conventional interpretation of the left quotation mark? You use a backslash to flag that you want to override the usual Stata interpretation.

```
. display "He said \`frog'."
He said `frog'.
```

It is perhaps unlikely, although clearly not impossible, that you do want this in practice. The problem is that you may appear to Stata to specify this, even though you are likely to do it purely by accident.

Suppose, for example, that you are looping over a series of datasets, reading each one into Stata, doing some work, and then moving on to the next.

Your code may look something like this:

```
. foreach f in a b c {
.     use "c:\data\this project\`f'"
.     and so on
. }
```

Do you see the difficulty now? Your intent is that the loop uses a local macro, which in turn takes the values `a`, `b`, and `c` to read in the datasets `a.dta`, `b.dta`, and `c.dta`. But Stata sees the last backslash as an instruction to escape the usual interpretation of the left quotation mark character that immediately follows. The resulting misunderstanding will crash your code.

Any Unix (including Macintosh) users reading this will feel smug, because they use the forward slash, /, within directory or filenames, and this problem never bites them. The way around the problem is by knowing that Stata will let you do that too, even under Windows. Stata takes it upon itself to translate between you and the operating system. You certainly need to use the forward slash in this example to escape the difficulty of the implied escape. Typing `use "c:\data\this project/‘f’"` would solve the problem. A backslash is only problematic whenever it can be interpreted as an escape character. In fact, you can mix forward and backward slashes willy-nilly in directory or filenames within Stata for Windows, so long as you do not use a backslash just before a left quotation mark.

The tidiest solution is to use forward slashes consistently, as in `use "c:/data/this project/‘f’"`, although admittedly this may clash strongly with your long-practiced Windows habits.

The Stata Journal (2008)
8, Number 3, pp. 448–449

Stata tip 66: ds—A hidden gem

Martin Weiss
University of Tuebingen
Tuebingen, Germany
martin.weiss@uni-tuebingen.de

ds is one of a few dozen "undocumented" commands in Stata whose names are available in help undocumented. Contrary to the help file assertion that "an undocumented command is a command of very limited interest, usually only to Stata programmers", ds is extremely helpful, both interactively and in programs. The main hurdle to its widespread adoption by Stata users seems to be limited awareness of its existence.

Stata users are generally familiar with the describe (see [D] **describe**) command. describe allows you to gain a rapid overview of the dataset in memory, or with the using modifier, a dataset residing on a hard disk or available on the Internet. ds also allows that with its detail option, omitting only the general information provided in the header of the output of the describe command. For example,

```
. sysuse uslifeexp2.dta
(U.S. life expectancy, 1900-1940)

. describe

Contains data from C:\Program Files\Stata10\ado\base/u/uslifeexp2.dta
  obs:            41                           U.S. life expectancy, 1900-1940
  vars:            2                           2 Apr 2007 14:39
  size:          574 (99.9% of memory free)    (_dta has notes)
--------------------------------------------------------------------------------
              storage   display    value
variable name   type    format     label      variable label
--------------------------------------------------------------------------------
year            int     %9.0g                  Year
le              float   %9.0g                  life expectancy
--------------------------------------------------------------------------------
Sorted by:  year

. ds, detail

              storage   display    value
variable name   type    format     label      variable label
--------------------------------------------------------------------------------
year            int     %9.0g                  Year
le              float   %9.0g                  life expectancy
--------------------------------------------------------------------------------
```

More importantly, ds also provides the means of identifying subsets of variables with specified properties. It complements lookfor (see [D] **lookfor**), which allows you to search for certain strings in variable names or variable labels. ds enhances this functionality extensively, letting you specify

- certain types of variables;

- whether variable labels, value labels, and characteristics have been attached, and if so, whether they match certain patterns; and

- variables with specific formats.

A further systematic feature is that you can specify either the subset of variables satisfying particular properties or the complementary subset that does not satisfy those properties. As a simple example of the latter, when you are using the `auto.dta`, the command `ds make, not` specifies all variables other than `make`.

These capabilities prove particularly handy with large or poorly known datasets. As a simple example, pretend you were not familiar with the `auto` dataset and were looking for string variables.

```
. sysuse auto.dta
(1978 Automobile Data)
. * show all variables featuring type string
. ds, has(type string)
make
```

While `describe` would list the variable types, leaving the task of finding a certain type to you, `ds` can provide precisely what you were looking for. Despite its "undocumented" status, a dialog box can ease navigation through the intricacies of this command: to try the dialog box, type `db ds`. Beyond the results shown as usual, `ds` also leaves behind a list of variables found in `r(varlist)`, which is available for use by subsequent commands, such as `list` (see [D] **list**) or `summarize` (see [R] **summarize**). Many of the properties that you can search for with `ds` can also be extracted with extended macro functions; see [P] **macro**. To illustrate, consider the `voter` dataset shipped with Stata.

```
. sysuse voter.dta
. * show all variables with value labels attached
. ds, has(vall)
candidat  inc
. * show all variables not of type float
. ds, not(type float)
candidat  inc       pfrac      pop
. * show mean of all variables with format %10.0g
. ds, has(format %10.0g)
pfrac  pop
. tabstat 'r(varlist)'
    stats |     pfrac        pop
----------+---------------------
     mean | 6.733333   104299.3
```

The help file accessed by typing `help ds` gives several more examples. The help file accessed by typing `help varfind_pat_examp` explains the use of wildcards within the specification of patterns.

The Stata Journal (2008)
8, Number 3, pp. 450–451

Stata tip 67: J() now has greater replicating powers

Nicholas J. Cox
Department of Geography
Durham University
Durham City, UK
n.j.cox@durham.ac.uk

The Mata standard function J() was generalized in the Stata update of 25 February 2008. This tip flags its greater replicating powers. Note that J() in Stata's original matrix language remains as it was.

Users of Mata will have become accustomed to the role of J() in creating matrices of constants. For example, once within Mata,

```
: J(5,5,1)
[symmetric]
        1   2   3   4   5

    1   1
    2   1   1
    3   1   1   1
    4   1   1   1   1
    5   1   1   1   1   1
```

You may be used to thinking of the way J() works like this: I want a 5×5 matrix, all of whose elements are the scalar 1. Another way of thinking about it is this: Give me 5 replicates or copies rowwise and 5 copies columnwise of the scalar 1. The results are identical when scalars are being replicated.

The second way of thinking about it helps in understanding the generalization now in place. What is to be replicated can now be a matrix, naturally including not only scalars but also vectors as special cases.

The help file gives full technical details and a variety of examples, but here is another. My Speaking Stata column in this issue (Cox 2008) mentions the bias on Fisher's z scale when estimating correlation r from sample size n of $2r/(n-1)$. The question is thus how big this is for a variety of values of r and n. We can quickly get a table from Mata:

```
: r = J(5, 1, (.1, .3, .5, .7, .9))
: r
        1    2    3    4    5

    1   .1   .3   .5   .7   .9
    2   .1   .3   .5   .7   .9
    3   .1   .3   .5   .7   .9
    4   .1   .3   .5   .7   .9
    5   .1   .3   .5   .7   .9
```

```
: n = J(1, 5, (10, 20, 50, 100, 200)')
: n
               1      2      3      4      5

    1         10     10     10     10     10
    2         20     20     20     20     20
    3         50     50     50     50     50
    4        100    100    100    100    100
    5        200    200    200    200    200

: 2 * r :/ (n :- 1)
                  1              2              3              4              5

    1   .0222222222    .0666666667    .1111111111    .1555555556             .2
    2   .0105263158    .0315789474    .0526315789    .0736842105    .0947368421
    3   .0040816327     .012244898    .0204081633    .0285714286    .0367346939
    4    .002020202    .0060606061    .0101010101    .0141414141    .0181818182
    5   .0010050251    .0030150754    .0050251256    .0070351759    .0090452261
```

Notice again how the first two arguments of J() are the numbers of replicates or copies, rowwise and columnwise, and not necessarily the numbers of rows and columns in the resulting matrix.

Reference

Cox, N. J. 2008. Speaking Stata: Correlation with confidence, or Fisher's z revisited. *Stata Journal* 8: 413–439.

The Stata Journal (2008)
8, Number 4, pp. 583–585

Stata tip 69: Producing log files based on successful interactive commands

Alan R. Riley
StataCorp
College Station, TX
ariley@stata.com

So, your interactive Stata session went well and you got some good results. Naturally, you made sure you kept a log file by using the `log` command ([R] **log**).

But, almost inevitably, you also made some errors in your commands. And perhaps you also have within your log some digressions, repetitions, or things that turned out to be not so interesting or useful. How do you now produce a log file based only on the successful commands? More importantly, how do you save the sequence of commands you issued so that you can reproduce your results?

Such questions are longstanding, and there are possible software solutions on various levels. You might reach for the Stata Do-file Editor or your favorite text editor or scripting language, or you might write a program using Stata commands such as `file` ([P] **file**) to edit the log file down to the valuable part. See, for example, Cox (1994) or Eng (2007) for some detailed suggestions.

Here I concentrate on two approaches that should help.

The first approach makes a good start by using Stata's Review window, which displays a history of commands submitted to Stata. The following steps will save a do-file consisting of all the interactive commands that did not result in an error.

There are three columns in the Review window: sequence number (the order in which commands were submitted to Stata), command (the command itself), and return code (the return code, or `_rc`, from the command if it exited with an error; this column is empty if the command completed successfully).

After issuing several commands to Stata interactively, some of which might have resulted in errors, click on the top of the return code (`_rc`) column in the Review window. This will sort the commands in the Review window: now all the commands that did not result in an error are grouped together (and within that group they will be in the order in which they were submitted to Stata). Beneath them will be all the commands resulting in errors, sorted by the return code. The screenshots below show what the Review window might look like before and after doing this.

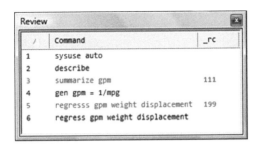
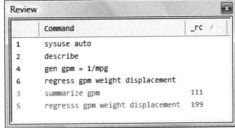

The first group is of interest; the second can be ignored. Select all the commands in the first group: click once on the first command in the group to select it, scroll until the last command in the group is visible, and then hold down the *Shift* key while clicking on that last command to select all the commands in the group.

Once all the valid commands have been selected, right-click anywhere in the Review window and choose **Save Selected...** to save those commands to a do-file, or choose **Send to Do-file Editor** to paste those commands into Stata's Do-file Editor.

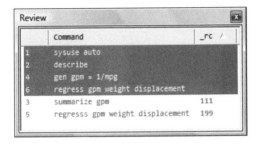
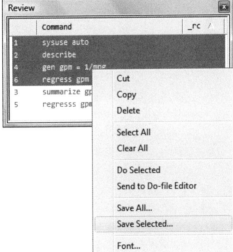

Naturally, this does not solve the questions of digressions, repetitions, or less worthwhile results. Nor is there an absolute guarantee that rerunning these commands would result in exactly the same state as you ended your session. There is a chance that some of the commands that produced an error changed your data, for example, if one of your commands was to run a do-file that stopped with an error after making certain changes to your data. However, knowing this approach should make your task easier.

The second approach uses `cmdlog`. Make sure that you run both `log` and `cmdlog` simultaneously. Then, with an eye on the log, edit the command log so that it contains only commands that were legal and useful. Then run the command log as a do-file to get the error-free log. In essence, there is less work to do that way.

References

Cox, N. J. 1994. os13: Using awk and fgrep for selective extraction from Stata log files. *Stata Technical Bulletin* 19: 15–17. Reprinted in *Stata Technical Bulletin Reprints*, vol. 4, pp. 78–80. College Station, TX: Stata Press.

Eng, J. 2007. File filtering in Stata: Handling complex data formats and navigating log files efficiently. *Stata Journal* 7: 98–105.

The Stata Journal (2008)
8, Number 4, pp. 586–587

Stata tip 70: Beware the evaluating equal sign

Nicholas J. Cox
Department of Geography
Durham University
Durham, UK
n.j.cox@durham.ac.uk

When you assign to a local or global macro the result of evaluating an expression, you can lose content because of limits on the length of string expression that Stata will handle. This tip explains this pitfall and the way around it in detail. The problem is described in [U] **18.3.4 Macros and expressions**, but it occurs often enough that another warning may be helpful. Illustrations will be in terms of local macros, but the warning applies equally to global macros.

What is the difference between the following two statements?

```
. local foo1 "just 23 characters long"
. local foo2 = "just 23 characters long"
```

Let's look at the results:

```
. display "`foo1'"
just 23 characters long
. display "`foo2'"
just 23 characters long
```

The answer is nothing, in terms of results. Do not let that deceive you. Two quite different processes are involved.

The first statement defines local macro `foo1` by copying the string `just 23 characters long` into it. The second statement does more. It first evaluates the string expression to the right of the equal sign and then defines the local macro by assigning the result of that evaluation to it.

The mantra to repeat is "the equal sign implies evaluation". Here the evaluation makes no difference to the result, but that will be not true in other cases. A more treacherous pitfall lies ahead.

It should be clear that we often need an evaluation. If the statements had been

```
. local bar1 lower("FROG")
. local bar2 = lower("FROG")
```

the results would have been quite different:

```
. display "`bar1'"
lower("FROG")
. display "`bar2'"
frog
```

The first just copies what is on the right to what is named on the left, the local macro `bar1`. The second does the evaluation, and then the result is put in the local macro.

Whenever you need an evaluation, you should be aware of a potential limit. `help limits` gives you information on limits applying to your Stata. As seen in the list, there are many different limits, and they can vary between Stata version and Stata flavor; there is no point in echoing that list here. But at the time of writing, the limit on the length of a string in a string expression in all flavors is 244 characters, which is quite a modest number.

The implication is that whenever you have a choice between copying and evaluation, always use copying. And if you must use evaluation, make sure that no single evaluation is affected by the limit of 244 characters. Further, be aware of work-arounds. Thus the `length()` function cannot report string lengths above 245 (yes, 245) characters, but the extended function : `length local` can measure much longer strings. See `help extended_fcn` for more details.

If you do not pay attention to the limit, you may run into problems. First, your local macro will be shorter than you want. Second, it may not even make sense to whatever you feed it to. Any error message you see will then not refer to the real underlying error, a truncated string. So your bug may be elusive.

For example, users sometimes collect variable names (or other names) in a list by using a loop centered on something like this:

```
. local mylist = "`mylist' `newitem'"
```

This may seem like self-evidently clear and correct code. But as `mylist` grows beyond 244 characters, the result of the evaluation entailed by = can only be to truncate the list, with possibly mysterious consequences.

The Stata Journal (2008)
8, Number 4, pp. 588–591

Stata tip 71: The problem of split identity, or how to group dyads

Nicholas J. Cox
Department of Geography
Durham University
Durham City, UK
n.j.cox@durham.ac.uk

Many researchers in various disciplines deal with dyadic data, including several who would not use that term. Consider couples in pairings or relationships of any kind, including spouses, lovers, trading partners, countries at war, teams or individuals in sporting encounters, twins, parents and children, owners and pets, firms in mergers or acquisitions, and so on. Dyads can be symmetric or asymmetric; benign, malign, or neutral; exclusive (one partner can be in only one dyad with their unique partner) or not. See Kenny, Kashy, and Cook (2006) for an introduction to the area from a social- and behavioral-science viewpoint.

Behind this intriguing variety lies a basic question: How can we handle dyad identifiers in Stata datasets? A natural data structure reflects the split identity of dyads: each dyad necessarily has two identifiers that researchers will usually read into two variables. This is indeed natural but also often poses a problem that we will need to fix. Suppose that Joanna and Jennifer are twins and that Billy Bob and Peggy Sue are twins. We might have observations looking like this:

```
. list person twin

     |    person         twin |
     |--------------------------|
  1. |    Joanna     Jennifer |
  2. |  Jennifer       Joanna |
  3. | Billy Bob    Peggy Sue |
  4. | Peggy Sue    Billy Bob |
```

And other variables would record data on each person. So the other variables for observation 1 could record the height, weight, number of children, etc., for Joanna, and those variables for observation 2 could record the same data for Jennifer.

In general, the identifiers need not be real names but could be any convenient string or numeric tags. Problems will arise if identifiers are not consistent across the two identifier variables. So with string identifiers, capitalization and other spelling must be identical, and all leading and trailing spaces should be trimmed. See Herrin and Poen (2008) for detailed advice on cleaning up string variables. Also, in general, there is no assumption so far that each person occurs just once in the dataset. Frequently, we will have multiple observations on each person in panel datasets, or we will have similar setups for other dyadic data.

If your dataset contains many hundreds or thousands of observations, you need automated methods for handling identifiers. Editing by hand is clearly time consuming, tedious, and error prone.

How do we spell out to Stata that Joanna and Jennifer are a pairing? Here is a simple trick that leads readily to others. We can agree that Joanna and Jennifer have a joint identity, which is (alphabetically) Jennifer Joanna. So we just need to sort those identifiers by observation, or rowwise. This can be done as follows:

```
. generate first = cond(person < twin, person, twin)
. generate second = cond(person < twin, twin, person)
. list person twin first second
```

	person	twin	first	second
1.	Joanna	Jennifer	Jennifer	Joanna
2.	Jennifer	Joanna	Jennifer	Joanna
3.	Billy Bob	Peggy Sue	Billy Bob	Peggy Sue
4.	Peggy Sue	Billy Bob	Billy Bob	Peggy Sue

This is all breathtakingly simple and obvious once you see the trick. You need to see that inequalities can be resolved for string arguments as well as for numeric arguments, and equally that `cond()` will readily produce string results if instructed.

Let us go through step by step. The `person < twin` string for Stata means that the value of `person` is less than the value of `twin`. For strings, *less than* means *earlier in alphanumeric order*. The precise order is that of `sort`, or of `char()`, not that of your dictionary. So `"a"` is less than `"b"`, but `"B"` is less than `"a"` because all uppercase letters are earlier in alphanumeric order than all lowercase letters. This precise order should only bite you if your identifiers are inconsistent, contrary to advice already given.

You might wonder quite how broad-minded Stata is in this territory. Do the functions `min()` and `max()` show the same generosity? No; `min("a", "b")` fails as a type mismatch.

The `cond()` function assigns results according to the answer to a question. See Kantor and Cox (2005) for a detailed introduction. If `person < twin`, then `first` takes on the values of `person` and `second` takes on the values of `twin`. If that is not true, then it is the other way around. Either way, `first` and `second` end up alphanumerically sorted.

I find it helpful to check through all the logical possibilities with an inequality, if only to reassure myself quickly that every possibility will produce the result I want. An inequality based on `<` will not be true if the operands satisfy `>` or if they satisfy `==`. Here, if the names are the wrong way around, they get swapped in the results for `person` or `twin`, which is as intended. What is easier to overlook is the boundary case of equality. If the names are the same, they will also be swapped, but that makes no difference; no harm is done and no information is lost. In the example of twins, names being the same might seem unlikely, but perhaps someone just used surnames, and the surnames are identical. As usual, however, data-entry errors are another matter. In some other examples of dyads, there may be good reason for the names to be consistently the same; if so, the problem discussed here does not arise at all.

An advantage of using `cond()` is that exactly the same code applies to numeric identifiers. Conversely, if the identifiers were numeric, it would be fine to code

```
. generate first = min(person, twin)
. generate second = max(person, twin)
```

and a quick check shows that this works even if the identifiers are identical.

The problem is now all but solved. We can group observations for each dyad with

```
. by first second: command
```

and if we need a unique identifier for each dyad—it will come in useful sooner or later— we can get that by typing

```
. egen id = group(first second)
```

The `egen, group()` command yields identifiers that are integers 1 and above. Note also its handy `label` option. For more detail on such variables, see Cox (2007).

Tips for dyads should come in pairs, so here is another. This is for the case in which there are precisely two observations for each dyad. Again twins are a clear-cut example. Often we will want to compare each twin with the other, say, by calculating a difference. Then the height difference for each twin is *this twin's height* minus *the other twin's height*, or

```
. by first second: generate diffheight = height - height[3 - _n]
```

Or if we had calculated that identifier variable mentioned earlier, the height difference would be

```
. by id: generate diffheight = height - height[3 - _n]
```

Where does the `[3 - _n]` subscript come from? Recall that under the aegis of `by:`, _n is determined *within* groups defined by the *byvarlist*, here, the identifier variable `id`. For more on that, see Cox (2002) or, perhaps more conveniently, the *Speaking Stata* column in this issue (Cox and Longton 2008). So _n will be 1 or 2. If it is 1, then $3 - 1$ is 2, and if it is 2, then $3 - 2$ is 1.

If that seems too tricky to recall, there are more commonplace ways to do it:

```
. by id: generate diffheight =
> cond(_n == 1, height - height[2], height - height[1])
```

or even

```
. by id: generate diffheight = height - height[2] if _n == 1
. by id: replace diffheight = height - height[1] if _n == 2
```

References

Cox, N. J. 2002. Speaking Stata: How to move step by: step. *Stata Journal* 2: 86–102.

———. 2007. Stata tip 52: Generating composite categorical variables. *Stata Journal* 7: 582–583.

Cox, N. J., and G. M. Longton. 2008. Speaking Stata: Distinct observations. *Stata Journal* 8: 557–568.

Herrin, J., and E. Poen. 2008. Stata tip 64: Cleaning up user-entered string variables. *Stata Journal* 8: 444–445.

Kantor, D., and N. J. Cox. 2005. Depending on conditions: A tutorial on the cond() function. *Stata Journal* 5: 413–420.

Kenny, D. A., D. A. Kashy, and W. L. Cook. 2006. *Dyadic Data Analysis*. New York: Guilford Press.

The Stata Journal (2008)
8, Number 4, pp. 592–593

Stata tip 72: Using the Graph Recorder to create a pseudograph scheme

Kevin Crow
StataCorp
College Station, TX
kcrow@stata.com

Following the update of 25 February 2008, the Graph Editor can now record a series of edits, name the recording, and apply the edits from the recording to other graphs. You can apply the recorded edits from the Graph Editor or from the command line. The edits can be applied from the command line when a graph is created, when it is used from disk, or whenever it is the active graph. See *Graph Recorder* in `help graph editor` for creating and playing recordings in the Graph Editor. For applying edits from the command line, see `help graph play` and the option `play(`*recordingname*`)` in `help std_options` and `help graph use`.

In this tip, I focus on the use of the Graph Recorder to create a graph scheme. A graph scheme specifies the overall look of the graph. If you want to create your own look for your graphs, you will want to create a scheme file. There is a problem with scheme files, however, because unless you know exactly how you want your scheme to be set up, creating a scheme can be very time consuming.

A shortcut to creating a scheme file is saving a graph recording to disk and replaying that recording on your graphs by using the `play()` option of the `graph` command. Using the Graph Recorder to create your graph recording also allows you to tinker with your graph's look on the fly without having to edit a scheme file. Let's walk through an example.

Suppose you want to create several graphs for a report and you want those graphs to have a specific look. To create your recording, you first need to draw the first graph of the report. Try

```
. sysuse auto
(1978 automobile data)
. scatter mpg weight
```

Now that your graph is in the Graph window, you can right-click on the window and select **Start Graph Editor** to start the Graph Editor. Next start the Graph Recorder by clicking on the **Start Recording** button, ●, so that you save your changes to memory. Once the graph looks the way you want, you then click on the same button, ● (which now has the tool tip **End Recording**), to save your changes to a `.grec` file. By default, Stata saves `.grec` files to your `PERSONAL/grec` directory.

If you want to tinker with the graph during a recording, but you do not want the changes to be saved to the `.grec` file, click on the **Pause** button, ▮▮, to temporarily stop saving the changes. To unpause the Recorder, click on the **Pause** button, ▮▮, again.

Now that you have a recorder file saved to disk, you can type your next `graph` command in your do-file or from the Command window and apply your scheme to the graph with the `play()` option. For example,

```
. scatter mpg turn, play("test.grec") saving(test1, replace)
```

Also, if you have graphs already created and saved to disk, you can apply your recording to those graphs by using the `play()` option of `graph use`. For example,

```
. graph use "oldfile.gph", play("test.grec") saving("new_file", replace)
```

The Stata Journal (2009)
9, Number 1, pp. 166–168

Stata tip 73: append with care![1]

Christopher F. Baum
Department of Economics
Boston College
Chestnut Hill, MA
baum@bc.edu

The `append` command is a useful tool for data management. Most users are aware that they should be careful when appending datasets in which variable names differ; for instance, `PRICE` in one dataset with `price` in another will lead to both variables appearing in different columns of the combined dataset. But one perhaps lesser-known feature of `append` is worth noting. What if the *names* of the variables in the two datasets are the same, but their *data types* differ? If that is the case, then the order in which you combine the datasets may matter and can even lead to different retained contents in the combined dataset. This is particularly dangerous (as I recently learned!) when a variable is held as numeric in one dataset and string in another.

Let's illustrate this feature with `auto.dta`. You may know that the `foreign` variable is a 0/1 indicator variable (0 for domestic, 1 for foreign) with a value label. Let's create two datasets from `auto.dta`: the first with only domestic cars (`autodom.dta`) and the second with only foreign cars (`autofor.dta`). In the former dataset, we will leave the `foreign` variable alone. It is numeric and will be zero for all observations. In the second dataset, we create a string variable named `foreign`, containing `foreign` for each observation.

```
. sysuse auto
(1978 Automobile Data)
. drop if foreign
(22 observations deleted)
. save autodom
file autodom.dta saved
. sysuse auto
(1978 Automobile Data)
. drop if !foreign
(52 observations deleted)
. rename foreign nondom
. generate str foreign = "foreign" if nondom
. save autofor
file autofor.dta saved
```

Let's pretend that we are unaware that the same variable name, `foreign`, has been used to represent numeric content in one dataset and string content in the other, and let's use `append` to combine them. We `use` the domestic dataset and `append` the foreign dataset:

---

1. Editors' note: As of Stata 11, `append` will no longer append string variables to numeric variables or vice versa unless you use the `force` option, which should only be used with extreme care. This tip still applies to users of previous versions of Stata.

```
. use autodom
(1978 Automobile Data)

. append using autofor
(note: foreign is str7 in using data but will be byte now)
(label origin already defined)

. describe foreign

                 storage  display    value
variable name    type     format     label      variable label

foreign          byte     %8.0g      origin     Car type

. codebook foreign
```

```
foreign                                                                   Car type

                  type:  numeric (byte)
                 label:  origin

                 range:  [0,0]                        units:  1
         unique values:  1                          missing .:  22/74

            tabulation:  Freq.   Numeric  Label
                            52         0  Domestic
                            22         .
```

Notice that **append** produced the following message:

```
(note: foreign is str7 in using data but will be byte now)
```

This message indicates that the `foreign` variable will be numeric in the combined dataset. The contents of the string variable in the using dataset have been lost, as you can see from the `codebook` output. Twenty-two cases are now classified as missing.

But quite a different outcome is forthcoming if we merely reverse the order of combining the datasets. It usually would not matter in what order we combined two or more datasets. After all, we could always use **sort** to place them in any desired order. But if we first **use** the foreign dataset and **append** the domestic dataset, we receive the following results:

```
. use autofor, clear
(1978 Automobile Data)

. append using autodom
(note: foreign is byte in using data but will be str7 now)
(label origin already defined)

. describe foreign

                 storage  display    value
variable name    type     format     label      variable label

foreign          str7     %9s
```

```
. codebook foreign
```

foreign	(unlabeled)

```
              type:  string (str7)
      unique values:  1                    missing "":  52/74
         tabulation:  Freq.  Value
                         52   ""
                         22   "foreign"
```

Again we receive the fairly innocuous message:

```
(note: foreign is byte in using data but will be str7 now)
```

Unfortunately, this message may not call your attention to what has happened in the append. Because the data type of the first dataset rules, the string variable is unchanged, and the numeric values in the using dataset are discarded. As the codebook shows, the variable foreign is now missing for all domestic cars.

You may be used to the notion, with commands like merge, that the choice of master and using datasets matters. You may not be as well aware that append's results may also be sensitive to the order in which files are appended. You should always take heed of the messages that append produces when data types are altered. If the append step changes only the data type of a numeric variable to allow for larger contents (for instance, byte to long or float to double) or extends the length of a string variable to allow for longer strings, no harm is done. But append does not highlight the instances, such as those we have displayed above, where combining string and numeric variables with the same name causes the total loss of one or the other's contents. In reality, that type of data type alteration deserves a warning message, or perhaps even an error. Until and unless such changes are made to this built-in Stata command, append with care.

The Stata Journal (2009)
9, Number 1, pp. 169–170

Stata tip 74: firstonly, a new option for tab2

Roberto G. Gutierrez
StataCorp
College Station, TX
rgutierrez@stata.com

Peter A. Lachenbruch[1]
Department of Public Health
Oregon State University
Corvallis, OR
peter.lachenbruch@oregonstate.edu

In many research contexts, we have several categorical variables and correspondingly want to look at several contingency tables. The `tab2` command allows us to do this. Using the `auto` dataset (after categorizing `mpg` to `mpgcat` and `price` to `cost`), we have

```
. sysuse auto, clear
(1978 Automobile Data)
. quietly egen mpgcat = cut(mpg), at(0, 15, 25, 35, 45) label
. quietly egen cost = cut(price), at(0, 4000, 6000, 10000, 16000) label
. tab2 foreign mpgcat cost rep78, chi2
-> tabulation of foreign by mpgcat
```

Car type	mpgcat 0-	15-	25-	35-	Total
Domestic	7	37	8	0	52
Foreign	1	10	8	3	22
Total	8	47	16	3	74

```
          Pearson chi2(3) =  12.9821   Pr = 0.005
```

(output omitted)

```
-> tabulation of cost by rep78
```

cost	Repair Record 1978 1	2	3	4	5	Total
0-	0	1	4	2	3	10
4000-	2	5	17	8	6	38
6000-	0	1	2	8	1	12
10000-	0	1	7	0	1	9
Total	2	8	30	18	11	69

```
          Pearson chi2(12) =  18.5482   Pr = 0.100
```

With 4 variables, we have 6 tables ($4 \times 3/2 = 6$). With more variables, the number of tables explodes quadratically. Even with 10 variables, we end up with 45 tables, which is likely to be more than really interests us. We may well want finer control.

Often there is one response variable of special interest. Our first focus may then be to relate that response to possible predictors. Suppose we wish to study if domestic or foreign cars differ on some variables. Thus we are interested in the three tables of `foreign` versus `mpgcat`, `cost`, and `rep78`. The `firstonly` option added to `tab2` in the update of 15 October 2008 allows us to get just the contingency table of the first-named variable versus the others.

---

1. Peter Lachenbruch's work was partially supported by a grant from the Cure JM Foundation.

```
. tab2 foreign mpgcat cost rep78, firstonly chi2
-> tabulation of foreign by mpgcat
                          mpgcat
  Car type |    0-      15-      25-      35- |    Total
-----------+--------------------------------+--------
  Domestic |     7       37        8        0 |       52
   Foreign |     1       10        8        3 |       22
-----------+--------------------------------+--------
     Total |     8       47       16        3 |       74
          Pearson chi2(3) =  12.9821   Pr = 0.005
-> tabulation of foreign by cost
                           cost
  Car type |    0-    4000-    6000-   10000- |    Total
-----------+--------------------------------+--------
  Domestic |     7       31        6        8 |       52
   Foreign |     4        9        7        2 |       22
-----------+--------------------------------+--------
     Total |    11       40       13       10 |       74
          Pearson chi2(3) =   5.3048   Pr = 0.151
-> tabulation of foreign by rep78
                    Repair Record 1978
  Car type |    1        2        3        4        5 |    Total
-----------+---------------------------------------+--------
  Domestic |    2        8       27        9        2 |       48
   Foreign |    0        0        3        9        9 |       21
-----------+---------------------------------------+--------
     Total |    2        8       30       18       11 |       69
          Pearson chi2(4) =  27.2640   Pr = 0.000
```

Here the number of tables is reduced from 6 to 3, a small change. However, for 10 variables (say, one response and nine predictors), the change is from 45 to 9.

This could have been programmed fairly easily with a `foreach` loop (see [P] **foreach**), but the new `firstonly` option makes life even a little easier.

The Stata Journal (2009)
9, Number 1, pp. 171–172

Stata tip 75: Setting up Stata for a presentation

Kevin Crow
StataCorp
College Station, TX
kcrow@stata.com

If you plan to use Stata in a presentation, you might consider changing a few settings so that Stata is easy for your audience to view. How you set up Stata for presenting will depend on several factors like the size and layout of the room, the length of the Stata commands you will issue, the datasets you will use, the resolution of the projector, etc. Changing the settings and saving those settings as a custom preference before you present can save you time and frustration. Also having a custom layout preference allows you to restore your setup should something happen in the middle of your presentation.

How you manipulate Stata's settings is platform dependent. This article assumes you are using Windows. If you use Stata for Macintosh or Unix, the advice is the same but the manipulations are slightly different.

First, make Stata's windows fill the screen. The maximize button is in the top right-hand corner of Stata (the maximize button is in the same place for all windows in Stata). After maximizing Stata, you will also want to maximize the Results window.

Once Stata is maximized, you will probably want to move the Command window. For most room layouts, you will want the Command window at the top of Stata so that your audience can see the commands you are typing. You achieve this by changing your windowing preferences to allow docking. In Stata, select **Edit** > **Preferences** > **General Preferences...**, and then select the **Windowing** tab in the dialog box that appears. Make sure that the check box for **Enable ability to dock, undock, or tab windows** is checked, and then click on the **OK** button. Next double-click on the blue title bar of the Command window and drag the window to the top docking button. Once the Command window is docked on top, it is a good idea to go back to the *General Preferences* dialog box and uncheck the box you changed. Doing this will ensure that your Command window stays at the top of Stata and does not accidentally undock.

Depending on the projector resolution, you will probably want to change the font, font style, and font size of the Command window. To change the font settings of a window in Stata, right-click within the window and select **Font...**. The font you choose is up to you, but we recommend Courier New as a serif font or Lucida Console as a sans serif font. You will also want to change the font size (14 is a good starting size) and change the font style to bold. Finally, we recommend that you resize the Command window so that you can see two lines (with the font and font size changed, you might find that long Stata commands do not fit on one line).

Once the Command window is set, you now want to change the font and font size of the Results window. After you have the font and font size selected, be sure that the line size in the Results window is at least 80 characters long to prevent wrapping of output. You can check your line size by typing the following command in Stata.

```
. display c(linesize)
```

Another setting to consider changing is the color scheme of the Results window from the default black background scheme to the white background scheme. To do this, bring up the *General Preferences* dialog box and, in the **Results color** tab, change the **Color scheme** drop-down box to **White background**. Switching to this color scheme will help people in the audience who are color-blind.

Next change the font and font size of the Review and Variables windows. For the Variables window, you might want to resize the Name, Label, Type, or Format columns depending on your dataset. For example, if your dataset has long variable names but does not have variable labels, you would want to drag the Name column wider in the Variables window. If you plan to use the Viewer, Graph window, Do-file Editor, or Data Editor in your presentation, you will probably also want to resize the window and change the font and font size to make them easier to view.

You can do far more advanced Stata layouts by enabling some windowing preferences in Stata. For example, if you would like more room in the Results window, you might consider pinning the Review and Variables windows to the side of Stata. Again bring up the *General Preferences* dialog box in Stata and go to the **Windowing** tab. Check the box labeled **Enable ability to pin or unpin windows** and then close the dialog. You should now see a pin button in the blue title bars of the Review and Variables windows. Clicking on this button makes the windows a tab on the left side of Stata. To view the windows, simply click on the tab.

Finally, save your settings as a preference. In Stata, select **Edit > Preferences > Manage Preferences > Save Preferences > New Preferences Set...**. A dialog box will prompt you to name your preference. To load this saved preference, select **Edit > Preferences > Manage Preferences > Load Preferences**, and then select your preference listed in the menu.

The Stata Journal (2009)
9, Number 2, pp. 321–326

Stata tip 76: Separating seasonal time series

Nicholas J. Cox
Department of Geography
Durham University
Durham, UK
n.j.cox@durham.ac.uk

Many researchers in various sciences deal with seasonally varying time series. The part rhythmic, part random character of much seasonal variation poses several graphical challenges for them. People usually want to see both the broad pattern and the fine structure of trends, seasonality, and any other components of variation. The very common practice of using just one plot versus date typically yields a saw-tooth or roller-coaster pattern as the seasons repeat. That method is often good for showing broad trends, but not so good for showing the details of seasonality. I reviewed several alternative graphical methods in a *Speaking Stata* column (Cox 2006). Here is yet another method, which is widely used in economics. Examples of this method can be found in Hylleberg (1986, 1992), Ghysels and Osborn (2001), and Franses and Paap (2004).

The main idea is remarkably simple: plot separate traces for each part of the year. Thus, for each series, there would be 2 traces for half-yearly data, 4 traces for quarterly data, 12 traces for monthly data, and so on. The idea seems unlikely to work well for finer subdivisions of the year, because there would be too many traces to compare. However, quarterly and monthly series in particular are so common in many fields that the idea deserves some exploration.

One of the examples in Franses and Paap (2004) concerns variations in an index of food and tobacco production for the United States for 1947–2000. I downloaded the data from http://people.few.eur.nl/paap/pbook.htm (this URL evidently supersedes those specified by Franses and Paap [2004, 12]) and named it `ftp`. For what follows, year and quarter variables are required, as well as a variable holding quarterly dates.

```
. egen year = seq(), from(1947) to(2000) block(4)
. egen quarter = seq(), to(4)
. gen date = yq(year, quarter)
. format date %tq
. tsset date
. gen growth = D1.ftp/ftp
```

Although a line plot is clearly possible, a scatterplot with marker labels is often worth trying first (figure 1). See an earlier tip by Cox (2005) for more examples.

```
. scatter growth year, ms(none) mla(quarter) mlabpos(0)
```

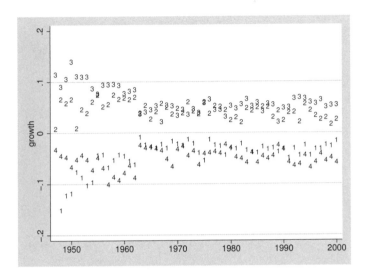

Figure 1: Year-on-year growth by quarter for food and tobacco production in the United States: separate series

Immediately, we see some intriguing features in the data. There seems to be a discontinuity in the early 1960s, which may reflect some change in the basis of calculating the index, rather than a structural shift in the economy or the climate. Note also that the style and the magnitude of seasonality change: look in detail at traces for quarters 1 and 4. No legend is needed for the graph, because the marker labels are self-explanatory. Compare this graph with the corresponding line plot given by Franses and Paap (2004, 15).

In contrast, only some of the same features are evident in more standard graphs. The traditional all-in-one line plot (figure 2) puts seasonality in context but is useless for studying detailed changes in its nature.

(*Continued on next page*)

```
. tsline ftp
```

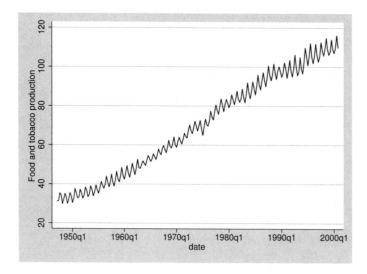

Figure 2: Quarterly food and tobacco production in the United States

The apparent discontinuity in the early 1960s is, however, clear in a plot of growth rate versus date (figure 3).

```
. tsline growth
```

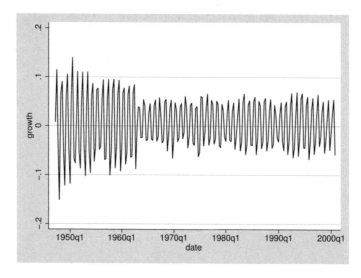

Figure 3: Year-on-year growth by quarter for food and tobacco production in the United States: combined series

An example with monthly data will push harder at the limits of this device. Grubb and Mason (2001) examined monthly data on air passengers in the United Kingdom for 1947–1999. The data can be found at http://people.bath.ac.uk/mascc/Grubb.TS; also see Chatfield (2004, 289–290). We will look at seasonality as expressed in monthly shares of annual totals (figure 4). The graph clearly shows how seasonality is steadily becoming more subdued.

```
. egen total = total(passengers), by(year)
. gen percent = 100 * passengers / total
. gen symbol = substr("1234567890ND", month, 1)
. scatter percent year, ms(none) mla(symbol) mlabpos(0) mlabsize(*.8) xtitle("")
> ytitle(% in each month) yla(5(5)15)
```

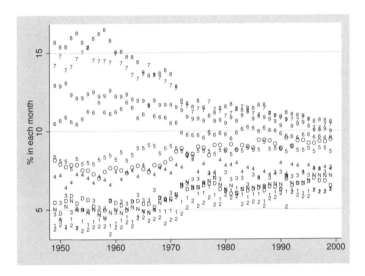

Figure 4: Monthly shares of UK air passengers, 1947–1999 (digits 1–9 indicate January–September; O, N, and D indicate October–December)

Because some users will undoubtedly want line plots, how is that to be done? The **separate** command is useful here: see Cox (2005), [D] **separate**, or the online help. Once we have separate variables, they can be used with the **line** command (figure 5).

(Continued on next page)

```
. separate percent, by(month) veryshortlabel
. line percent1-percent12 year, xtitle("") ytitle(% in each month) yla(5(5)15)
> legend(pos(3) col(1))
```

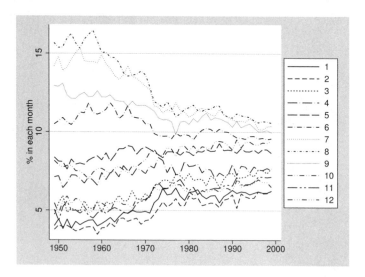

Figure 5: Monthly shares of UK air passengers, 1947–1999

You may think that the graph needs more work on the line patterns (and thus the legend), although perhaps now the scatterplot with marker labels seems a better possibility.

If graphs with 12 monthly traces seem too busy, one trick worth exploring is subdividing the year into two, three, or four parts and using separate panels in a by() option. Then each panel would have only six, four, or three traces.

References

Chatfield, C. 2004. *The Analysis of Time Series: An Introduction.* 6th ed. Boca Raton, FL: Chapman & Hall/CRC.

Cox, N. J. 2005. Stata tip 27: Classifying data points on scatter plots. *Stata Journal* 5: 604–606.

———. 2006. Speaking Stata: [b]Graphs for all seasons. *Stata Journal* 6: 397–419.

Franses, P. H., and R. Paap. 2004. *Periodic Time Series Models.* Oxford: Oxford University Press.

Ghysels, E., and D. R. Osborn. 2001. *The Econometric Analysis of Seasonal Time Series.* Cambridge: Cambridge University Press.

Grubb, H., and A. Mason. 2001. Long lead-time forecasting of UK air passengers by Holt–Winters methods with damped trend. *International Journal of Forecasting* 17: 71–82.

Hylleberg, S. 1986. *Seasonality in Regression*. Orlando, FL: Academic Press.

Hylleberg, S., ed. 1992. *Modelling Seasonality*. Oxford: Oxford University Press.

Subscriptions are available from StataCorp, 4905 Lakeway Drive, College Station, Texas 77845, telephone 979-696-4600 or 800-STATA-PC, fax 979-696-4601, or online at

http://www.stata.com/bookstore/sj.html

Back issues of the *Stata Journal* may be ordered online at

http://www.stata.com/bookstore/sjj.html

Individual articles three or more years old may be accessed online without charge. More recent articles may be ordered online.

http://www.stata-journal.com/archives.html

The *Stata Journal* is published quarterly by the Stata Press, College Station, Texas, USA.

Address changes should be sent to the *Stata Journal*, StataCorp, 4905 Lakeway Drive, College Station, TX 77845, USA, or emailed to sj@stata.com.